QUESTIONING EVOLUTION

An Interrogative Critique of Darwinian Assumptions

MARK BURTON

ISBN 10:   0-9744439-3-X
ISBN 13:   978-0-9744439-3-5

CONTENTS

Preface

1. **Is it possible to adequately address a topic using only the interrogative form?**
2. Would a lengthy list of questions be too tedious for the reader?
3. Or would a lengthy list be necessary to demonstrate that the author had given the issue at hand serious thought?
4. Would it help if the questions were not asked haphazardly but were arranged in a coherent manner?
5. Would it help to highlight key issues in bold lettering?
6. Would it help to categorize questions into different groups or chapters?
7. Could a series of specific questions be asked in a way that would convey broader implications "between the lines" or in the aggregate, allowing readers to infer unwritten conclusions?

8.   Would asking questions, rather than making declarative statements, be a less arrogant way of addressing some of life's great unresolved mysteries?

9.   Could Charles Darwin's "descent of man" theory of Evolution, for example, be addressed in such a way?

10.   Could a series of questions be used to expose underlying and often ignored assumptions of this theory that all life evolved from a single, original organism?

11.   Would such a list of questions be akin to a lengthy pop quiz for those who believe in Evolution?

12.   Would a large number of unanswerable questions raise doubts about the scientific validity of Evolution?

13.   If so, would that, in turn, raise doubts about the validity of atheist materialism (i.e., denial of the supernatural) and its underlying assumptions, by undermining the only theory the atheist materialists have for the existence of such a wide variety of life?

In The Beginning

14. **What is life?**

15. Is life, as one dictionary entry defines it, merely the property or quality manifested in functions such as metabolism, growth, response to stimuli and reproduction?

16. Or does life require some sort of natural or supernatural "spirit" to exist?

17. Do living things have to be organic or can they be inorganic?

18. Do living things require deoxyribonucleic acid (DNA)?

19. What's the difference between a living organism and inanimate matter?

20. What's the difference between a living organism and the physical remains of an organism that has died?

21. What is death?

22. Why does death occur?

23. In general, why does death exist as a phenomenon?

24.   That is, why does reality's *status quo* apparently include physical death as a default, rather than immortality?

25.   Are there different kinds of death?

26.   Is the death of a plant no different than the death of a human being?

27.   Is the death of a plant just a physical or natural phenomenon, devoid of anything spiritual or supernatural?

28.   Does the death of a human being include spiritual or supernatural phenomena, as well as physical or natural phenomena?

29.   What's the difference between (1) a living human being, and (2) a fresh human corpse hooked up to a respirator, an artificial heart, a feeding tube, a dialysis machine and other equipment that maintain the body's primary physical functions?

30.   Why do living organisms exist in the first place?

31.   **What is the origin of life?**

32.   If life did not originate from God or some other supernatural source, then how did life originate?

33.   Did nonlife produce life?

34.   If so, how?

35.   What was the first living organism?

36.   Was the first living organism single-celled?

37.   Or was it a multicellular organism?

38.   Or is it possible that it was something less than a cell?

39.   **When did life originate?**

40.   More than four billion years ago?

41.   Or less than 10,000 years ago?

42.   Or at some point in-between?

43.   Is it possible to know, with certainty, when life originated without first knowing where life originated?

44.   **Where did life originate?**

45.   Did life originate in one location only?

46.   Did the first living organism originate on Earth or somewhere else?

47.   Is there any evidence that life originated somewhere other than Earth?

48.   If life didn't originate on Earth, then how did one or more living organisms first come to exist on Earth?

49.   If life didn't originate on Earth, then when did one or more living organisms first come to exist on Earth?

50.   What were those organisms?

51.   Where did they come from?

52.   How did they originate?

53.   When they first arrived on Earth, were they single-celled organisms, multicellular organisms or something less than a cell?

54.   If life originated elsewhere and was somehow transferred to Earth, did only one organism get transferred to Earth?

55.   Or were multiple organisms transferred to Earth, either all at once or separately over time (e.g., from comet or asteroid collisions)?

56.   Did life originate in multiple locations separately?

57.   For example, did life originate on Earth independently of any life that might exist elsewhere in the universe?

58.   If the first living organism on Earth originated on Earth, was life on Earth later supplemented or altered by the arrival of other organisms that originated elsewhere in the universe?

59.   If so, when and how often did this occur?

60.   For each such occurrence, how was life on Earth affected?

61.   If life has never been transferred to Earth from elsewhere, then *where* on Earth did the first living organism on Earth originate?

62.   Did life on Earth originate near a volcanic vent beneath the ocean?

63.   Or did life on Earth originate in the ice near one of the poles?

64.   Or did life on Earth originate in a freshwater pool of primordial ooze?

65.   Is there any way to know for sure where the first living organism originated?

66.   If life has never been transferred to Earth from elsewhere, then *when* did the first living organism on Earth originate?

67.   Is there any way to know for sure when the first living organism originated?

68.   If life has never been transferred to Earth from elsewhere, then *what* was the first living organism on Earth?

69.   Was it a single-celled organism, a multicellular organism or something less than a cell?

70.   Is there any way to know for sure?

71.   How could something less than a cell, such as a virus, survive and reproduce on its own?

72.   Or would viruses, or anything less than a cell, not be considered life?

73.   If life has never been transferred to Earth from elsewhere, then *how* did such a variety of life come to exist on Earth?

BUILDING BLOCKS

74. Did all life on Earth today evolve from a single, primitive, original organism that originated on Earth, as Darwin himself suggested in Chapter XV of *The Origin of Species by Means of Natural Selection* when he wrote that "...all the organic beings that have ever lived on this earth may be descended from some one primordial form"?[1]

75. If so, how did that first organism originate?

76. If the cell is the fundamental structural unit of all known living organisms and all such organisms evolved from that single, original organism, does that mean the original organism consisted of at least one cell?

77. When did the very first cell originate?

78. Where did the very first cell originate?

---

[1] Some sources indicate that the full title of Darwin's *Origin of Species* is *The Origin of Species by Means of Natural Selection, or the Preservation of Favoured Races in the Struggle for Life.*

79.   How did the very first cell originate?

80.   How many components did the very first cell have?

81.   What were those components?

82.   For example, did the first cell have DNA, cytoplasm and an outer membrane?

83.   Did the first cell have a nucleus?

84.   Did the first cell have mitochondria, chloroplasts, ribosomes or messenger ribonucleic acid (RNA)?

85.   Did the first cell have any components that are *not* found in bacteria, plant or animal cells today?

86.   When, where and how did each of the various components of the first cell originate?

87.   Did the components of the first cell originate simultaneously?

88.   If not, how long did it take for all of the components of the first cell to originate?

89.   When did those components first begin to function interactively in a manner sufficient for the first cell to be considered alive?

90.   What caused those components to begin functioning interactively?

91.   What functions did the components carry out?

92.   Did the first cell successfully replicate?

93.   If so, when did the first cell replicate?

94.   Where did the first cell replicate?

95.   How did the first cell replicate?

96.   What caused the first cell to replicate?

97.   When did DNA originate?

98.   Where did DNA originate?

99.   How did DNA originate?

100.   What were the substances or components that made up the very first DNA?

101.   When did each of the substances or components that made up the first DNA come into existence?

102.   Where did each of the substances or components that made up the first DNA come into existence?

103.   How did each of the substances or components that made up the first DNA come into existence?

104.   Were those substances or components originally formed by random chance combinations of organic molecules?

105.   If so, which organic molecules randomly combined to form each DNA predecessor substance or component?

106.   How long did it take for each DNA predecessor substance or component to come into existence?

107.   How long did it take those DNA predecessor substances or components to form the first DNA?

108.   How did the information, or instructions, found in DNA for cell replication, protein production and other functions first come into existence?

109.   Which came first, the cell or DNA?

110.   Would a cell without DNA still be considered a living organism?

111.   Did DNA originate separately from the first cell?

112.   If so, how did DNA come to be an integral part of cells?

113.   Why, though, would DNA have instructions for cell replication and other functions if DNA originated separately from cells?

114.   Did DNA, instead, originate inside of the first cell after the cell formed?

115.   If so, how long did it take for DNA to come into existence after the first cell had formed?

116.   Did the first cell and DNA originate together, simultaneously, along with various other cell components?

117.   If so, what are the odds of that happening, given that the substances that make up DNA are very different than the proteins and other substances that make up the cell's outer membrane and cytoplasm?

118.   Given the importance of proteins in cells, which types of proteins were part of the very first cell?

119.   How many types of proteins already existed on Earth when the first cell originated?

120.   Which, if any, of those types of proteins were *not* involved in the formation of the first cell?

121.   Why were they not involved?

122.   When the first cell came into existence, did only a small number of proteins exist in a single location?

123.   Or did large numbers of proteins exist in multiple locations throughout the world before the first cell came into existence?

124.   If large numbers of proteins existed in multiple locations throughout the world, did they randomly combine with other substances only once at one location to form the first cell?

125.   Or did more than one cell originate independently via such random chance combinations at one location?

126.   Or did more than one cell originate independently at multiple locations?

127.   How long had the pre-cell proteins existed before some or all of them combined with other substances into the first cell?

128.   What specific environmental conditions or circumstances led to the random chance combination of proteins and other substances into the first cell?

129.   Prior to the origination of the first cell, did any random chance combinations of proteins occur that did *not* result in a cell?

130.   If so, how many such combinations occurred?

131.   When did each occur?

132.   Where did each occur?

133.   In other words, was the first cell or the first living organism the result of the very first random chance combination of proteins and other substances?

134.   Or had proteins and other substances been randomly combining over and over again until, eventually, the first cell or first living organism was formed?

135.   What was the first type of pre-cell protein to come into existence?

136.   When did the first pre-cell protein originate?

137.   Where did the first pre-cell protein originate?

138.   How did the first pre-cell protein originate?

139.   If different types of proteins existed before the first cell came into existence, then when, where and how did each come into existence?

140.   **For example, did one or more pre-cell proteins originate from random chance combinations of amino acids?**

141.   If so, how many protein-forming, random chance combinations of amino acids occurred prior to the formation of the first cell?

142.   Which types of proteins and how many of each originated by random chance combinations of amino acids?

143.   What specific environmental conditions or circumstances led to the random chance combination of amino acids into the very first protein?

144.   Prior to the formation of the very first protein, did any random chance combinations of amino acids occur that did *not* form a protein?

145.   If so, how many such combinations occurred?

146.   When did each such combination occur?

147.   Where did each such combination occur?

148.   In other words, was the very first protein the result of the very first random chance combination of amino acids?

149.   Or had amino acids been randomly combining over and over again until, eventually, the first protein was formed?

150.   How long would a pre-cell protein exist?

151.   Did pre-cell proteins ever decay or get destroyed by external forces?

152.   If the first cell originated in part from a random chance combination of proteins that, in turn, originated from random chance combinations of amino acids, then what was the first amino acid?

153.   When did the first amino acid come into existence?

154.   Where did the first amino acid come into existence?

155.   How did the first amino acid come into existence?

156.   Was the first amino acid formed by random chance combinations of organic molecules or other substances?

157.   If so, how many amino acid forming, random chance combinations of organic molecules or other substances occurred prior to the formation of the first protein?

158.   Which types of amino acids and how many of each originated by such random chance combinations?

159.   Which types of amino acids and how many of each existed prior to the formation of the very first protein?

160.   When the first protein came into existence, did only a small number of amino acids exist in a single location?

161.   Or did large numbers of amino acids exist in multiple locations throughout the world before the first protein came into existence?

162.   If large numbers of amino acids existed in multiple locations throughout the world, did they randomly combine only once at one location to form the first pre-cell protein?

163.   Or did more than one pre-cell protein originate independently from such combinations at one location?

164.   Or did more than one pre-cell protein originate independently at multiple locations?

165.   Which types of amino acids and how many of each randomly combined to form the first protein?

166.   How long had amino acids existed prior to the formation of the very first protein?

167.   Before proteins existed, how long would a single amino acid exist?

168.   Did pre-protein amino acids ever decay or get destroyed by external forces?

169.   If the first cell originated in part from (1) random chance formations of DNA and other components which, in turn, randomly combined with (2) a random chance combination of proteins that, in turn, originated from random chance combinations

of amino acids; all of which ultimately originated from random chance combinations of organic molecules or other substances, then what types of organic molecules or substances were involved?

170. For example, what types of organic molecules or other substances randomly combined to form the first amino acid on Earth?

171. Did the first amino acid originate from random chance combinations of methane, ammonia, water vapor and possibly other substances, as some have suggested, with or without the effects of lightning or sunlight?

172. If so, when, where and how did methane first come to exist on Earth or in Earth's atmosphere?

173. When, where and how did ammonia first come to exist on Earth or in Earth's atmosphere?

174. When, where and how did water vapor first come to exist on Earth or in Earth's atmosphere?

175. If other substances or organic molecules were involved in the formation of the first amino acid, then when, where and how did each first come to exist on Earth or in Earth's atmosphere?

176. How long did it take each of those substances or organic molecules to come into existence?

177. How long had each of those substances or organic molecules been on Earth or in Earth's atmosphere prior to the formation of the first amino acid?

178.   Prior to the formations of the first DNA component and the first amino acid, did any random chance combinations of organic molecules or other substances occur that did *not* form a DNA component or an amino acid?

179.   If so, how many such combinations occurred?

180.   When did each such combination occur?

181.   Where did each such combination occur?

182.   What did each such combination produce?

183.   In other words, was the first DNA or DNA component the result of the very first random combination of organic molecules or other substances?

184.   Or had organic molecules or other substances been randomly combining over and over again until, eventually, the first DNA or DNA component was formed?

185.   Was the very first amino acid the result of the very first random combination of organic molecules or other substances?

186.   Or had organic molecules or other substances been randomly combining over and over again until, eventually, the first amino acid was formed?

187.   Which came first, DNA or amino acids?

188.   **Could life exist without organic molecules and the things that make organic molecules possible?**

189.   If so, how?

190.   Which came first, organic molecules or inorganic molecules?

191.   What was the first organic molecule to come into existence?

192.   When did the first organic molecule come into existence?

193.   Where did the first organic molecule come into existence?

194.   How did the first organic molecule come into existence?

195.   What was the first inorganic molecule to come into existence?

196.   When did the first inorganic molecule come into existence?

197.   Where did the first inorganic molecule come into existence?

198.   How did the first inorganic molecule come into existence?

199.   Did the first organic molecule originate from random chance combinations of atoms?

200.   Did the first inorganic molecule originate from random chance combinations of atoms?

201.   What was the first atom to come into existence?

202.   When did the first atom come into existence?

203.   Where did the first atom come into existence?

204.   How did the first atom come into existence?

205.   How many different types of atoms exist today?

206.   If there are really more than 100 different types of atoms, why are there so many?

207.   After the first type of atom came into existence, when did each subsequent type of atom first come into existence?

208.   Where did each subsequent type of atom come into existence?

209.   How did each subsequent type of atom come into existence?

210.   If atoms are made up of subatomic particles, then when did those subatomic particles first come into existence?

211.   Where did subatomic particles first come into existence?

212.   How did subatomic particles first come into existence?

213.   If some subatomic particles, such as protons and neutrons, are made up of quarks and possibly gluons, then when did quarks and gluons first come into existence?

214.   Where did quarks and gluons first come into existence?

215.   How did quarks and gluons first come into existence?

216.   Do quarks and gluons, in turn, have subcomponents?

217.   If so, what are those subcomponents?

218.   Does it ever end?

219.   That is, does some truly elementary particle, string, membrane or other component of matter exist that has no subcomponent(s) of its own?

220.   If so, what is it?

221.   If not, how can that be?

222.   In general, why is matter so complex?

223.   For example, if some atoms are radioactive, why?

224.   Why would radioactivity even be possible?

225.   If there are different types of radioactive decay, such alpha, beta and gamma, did they all come into existence simultaneously?

226.   If not, which came into existence first?

227.   Which came into existence second?

228.   Why would different types of radioactive decay exist?

229.   Are there any types of radioactive decay that are currently unknown?

230.   If so, what are they?

231.   If not, then why aren't there more types of radioactive decay?

232.   If heavier atoms are formed in nature when stars collapse, then when did the first stars come into existence?

233.   Where did the first stars come into existence?

234.   How did the first stars come into existence?

235.   If stars are formed by condensed pockets of hydrogen gas, then when, where and how did hydrogen first come into existence?

236.   When did the energy, matter and physical forces that make stars, atoms and molecules possible, come into existence?

237.   Where did the energy, matter and physical forces that make stars, atoms and molecules possible, come into existence?

238.   How did the energy, matter and physical forces that make stars, atoms and molecules possible, come into existence?

239.   Did those things, or some predecessor aspect of the space-time continuum, come into existence in a "Big Bang"?

240.   That is, did the entire universe just pop into existence out of nothing, for no reason whatsoever?

241.   In other words, as some have phrased it, did nothing produce everything?

242. If science is the study of cause and effect through experimentation and observation, is it scientific to abandon the notion of cause and effect in postulating that the entire universe popped into existence out of nothing for no reason whatsoever?

243. Furthermore, if nothing existed before the Big Bang, why was matter not evenly distributed afterward?

244. That is, why were there concentrations of matter that coalesced over time, eventually forming stars and planets, without which life as we know it presumably couldn't have come into existence?

245. Has the universe, instead, existed eternally, without a Big Bang or any other origin?

246. Or does entropy preclude the possibility of an eternal universe (barring some as-yet-unknown phenomenon that counteracts entropy)?

247. That is, would the universe have "fizzled out" by now if it were eternal (i.e., without origin)?

248. If the universe is eternal, would that mean that Earth, and perhaps life on Earth, have existed eternally, ruling out any need to explain the origin of life?

249. Or does the existence of radioactive material on Earth preclude the possibility that the planet is eternal (barring the possibility of some sort of continuous resupply of radioactive material from an extraterrestrial source)?

250.   That is, wouldn't all radioactive material on Earth have fully decayed by now, if Earth were eternal?

251.   How old, then, is the Earth, if it's not eternal?

252.   Is there any way to accurately determine the age of the Earth?

253.   Does the angular velocity of the Earth's movement around the sun matter?

254.   Has the angular velocity of the Earth around the sun always been about 365 24-hour days?

255.   Or has the angular velocity of the Earth around the sun changed significantly at one or more points in the past, perhaps as a result of an asteroid, comet or other type of collision with the Earth?

256.   Has the distance between the Earth and the sun always been about 93 million miles?

257.   Or has that distance changed significantly at one or more points in the past, perhaps as a result of an asteroid, comet or other type of collision with the Earth?

258.   Has the size of the Earth ever changed since the Earth first coalesced into a planet?

259.   For example, has the Earth's diameter ever increased, perhaps from an expanding crust, leading to continental drift?

260.   Has the angular velocity of the Earth's rotation about its axis always been about 24 hours?

261.   Or has the angular velocity of the Earth's rotation about its axis changed significantly at one or more

points in the past, perhaps as a result of a change in Earth's diameter, or because of an asteroid, comet or other type of collision with the Earth?

262.   Are the age of the Earth, or its velocity around the sun, or its distance from the sun, or its size, or its rotational velocity about its axis, important factors in determining the validity of the "descent of man" theory?

263.   For example, how old does the Earth have to be for the "descent of man" theory to work?

264.   How much time did it take for the Earth to form?

265.   How, indeed, did so many different atoms and so many different molecules find their way to this one particular spot in the universe, and come together in just the right way to form a planet capable of sustaining life?

266.   How much time did it take for the organic molecules and other substances to form that are necessary for amino acids, proteins and DNA to exist?

267.   If there were any significant changes to the Earth's size, orbit or movements in the past, did such changes have any effect(s) on the development of organic molecules, amino acids, proteins, DNA or living organisms?

268.   If so, what effect(s)?

269.   For example, were any of the processes that led to the formation of organic molecules, amino acids, proteins, DNA or living organisms ever disrupted by

an asteroid, comet or other type of collision with the Earth?

270.   If so, how many times did such disruptions occur?

271.   When did such disruptions occur?

272.   How much time did it take for the reestablishment of conditions suitable for life after each disruption?

273.   And just out of curiosity, how old is the universe, if it's not eternal?

274.   Are energy, matter and physical forces as old as the universe itself?

275.   In any case, why do energy, matter and physical forces (such as gravity and electromagnetism) each act in the way that they do, rather than in some other way?

276.   Why do energy, matter and physical forces exist in the first place?

277.   More specifically, why do energy, matter and physical forces exist and act in ways that allow the existence of life under certain circumstances?

278.   In addition, does the universe include "fundamental constants" which, if one or more of them had been different, would have precluded the possibility of life?

279.   If so, does the existence of such constants suggest that the universe is less likely to be a product of random chance and more likely to be the result of an intelligent design?

280. Is it fair, however, to include such questions about the origin, variation and interactions of nonliving things when examining the various theories of Evolution, given that Evolution is only intended to explain the variety of living organisms?

281. **Or is it essential to understand that the various theories of Evolution fail to fully explain the origin, variation and interactions of nonliving things—including the very proteins, amino acids, organic molecules, atoms, subatomic particles, physical laws and fundamental constants that are indispensable to Evolutionary assumptions about the origin and variety of life?**

282. If Evolutionists don't know what the first living organism was, or when it originated, or where it originated, or how it originated, isn't it fair to say that Evolutionists are merely *assuming* that all life evolved from a single, original, living organism?

283. Whether that assumption is true or not, how many random chance combinations of organic molecules, amino acids, proteins and presumably other substances occurred, in total, prior to the formation of the first living organism on Earth?

284. In total, how long did it take for those random chance combinations to occur before the first living organism came into existence?

285.   Is the Earth old enough for so many random chance combinations to have occurred?

286.   Even if the Earth is technically old enough, did suitable conditions on Earth exist long enough for such random chance combinations to have occurred?

287.   What are the odds that such a series of random chance combinations would not only occur, but would result in a living organism?

DESCENT

288.   **Is there any scientific reason to assume that all life on Earth evolved from the first living organism on Earth?**

289.   After all, if the first living organism on Earth could come into existence from nonliving matter, couldn't *other* organisms also come into existence in a similar manner, independently of the first one?

290.   Didn't Darwin himself mention this possibility in Chapter XV of *The Origin of Species* when he stated that "no doubt it is possible, as Mr. G. H. Lewes as urged, that at the first commencement of life many forms were evolved (*sic*)"?

291.   In other words, if random chance combinations of proteins and other substances formed the first living organism on Earth three or four billion years ago, why should it stop there?

292.   If it only took a billion or so years since Earth's formation for random chance combinations of proteins and other substances to form the first living

organism, wouldn't it be likely that random chance combinations of proteins and other substances would continue to produce new organisms over the next few billion years, independently of any previously existing organisms?

293.   Is there any evidence of that?

294.   If not, why not?

295.   If so, how many organisms have come into existence independently of each other (i.e., out of nonliving matter)?

296.   How many, if any, were single-celled?

297.   How many, if any, were multicellular?

298.   How many, if any, were something less than a cell?

299.   When did each one come into existence?

300.   Where did each one come into existence?

301.   How did each one come into existence?

302.   Have any such organisms come into existence recently?

303.   If so, what has independently come into existence recently?

304.   If not, why not?

305.   Regardless of when they came into existence, were all organisms of independent origin the same species?

306.   Or did different species originate independently of each other?

307.   Did each of the earliest organisms of independent origin have the ability to reproduce?

308.   Or did one or more of the earliest living organisms of independent origin lack the ability to reproduce and die out?

309.   Did all of the earliest organisms of independent origin that had the ability to reproduce actually do so, or did some of them die out before reproducing?

310.   Which ones died before reproducing and which ones survived and reproduced?

311.   For example, did the very first organism to come into existence survive and reproduce, or did it die before reproducing?

312.   Did a second organism of independent origin succeed in reproducing or did it, too, die out?

313.   How many organisms of independent origin survived long enough to successfully reproduce and how many did not?

314.   Of these, which was the very first to reproduce?

315.   When did it first reproduce?

316.   Where did it first reproduce?

317.   How did it first reproduce?

318.   Was, or were, the very first offspring single-celled, multicellular or something less than a cell?

319.   Was, or were, the first offspring the same type of organism as the parent organism, or something different?

320.   If the first offspring was or were different from the parent organism, what was the difference and what caused the difference?

321.   Did the first offspring have the ability to reproduce?

322.   If so, did the first offspring survive long enough to reproduce?

323.   If so, when did the first offspring reproduce?

324.   Where did the first offspring reproduce?

325.   How did the first offspring reproduce?

326.   What was the result of such reproduction?

327.   Was that third generation substantially different from the first or second?

328.   If so, why and in what way?

329.   **Of all the original, independent organisms that successfully reproduced, how many, if any, produced mutated offspring or mutated descendants in later generations?**

330.   For example, did the first organism to successfully reproduce ever have mutated offspring or mutated descendants in later generations?

331.   Or did the first such successful organism survive and reproduce without producing any mutated offspring or mutated descendants?

332.   If so, how many non-mutated generations descended from that first successful organism?

333.   Do any non-mutated descendants of any original, independent organisms still exist today?

334.   If so, which species still exist that have descended from one or more original, independent organisms without ever mutating?

335. Or did such descendants become extinct at some point?

336. If any became extinct, when did such extinction(s) occur?

337. Where did such extinction(s) occur?

338. Why did such extinction(s) occur?

339. How many organisms of independent origination were in existence when the first mutation occurred?

340. Which one had the very first mutated offspring or mutated descendant?

341. In other words, for the "descent of man" theory to be true, random chance physical mutations are required, so wouldn't the very first such mutation have to have been a mutation of: (1) the very first organism to come into existence; (2) some other original organism that came into existence after, but independently of, the very first organism; or (3) one of the descendants of either the very first organism or some other original organism that came into existence after, but independently of, the very first organism?

342. Did the first mutation occur in the first generation of offspring of the first organism to produce a mutated descendant?

343. If not, how many generations passed before the first mutation occurred?

344.   How much time passed between the formation of the first organism to produce a mutated descendant and the first mutation?

345.   When did the first mutation occur?

346.   More than four billion years ago?

347.   Or less than 10,000 years ago?

348.   Or at some point in-between?

349.   How much time passed between the formation of the very first living organism and the first mutation (regardless of which organism was the first to produce mutated descendants)?

350.   Where did the first mutation occur?

351.   What caused the first mutation?

352.   For example, was the first mutation caused by inherently error-prone DNA?

353.   If so, why was the DNA prone to errors?

354.   If not, was the first mutation caused by some external influence, such as exposure to certain chemicals, cosmic radiation or radioactive material on Earth?

355.   Or was the first mutation caused by something else?

356.   If so, what?

357.   What was the result of the first mutation?

358.   Was the first mutated organism a different species than its parent(s)?

359.   Or was it just a slight variation of its parent(s)?

360. Was the first mutated organism a single-celled organism, a multicellular organism or something less than a cell?

361. Was the first mutation successful?

362. That is, did the first mutated organism survive and reproduce, or did it die without reproducing?

363. How many unsuccessful mutations occurred, if any, prior to the first successful mutation?

364. In total, how many times did the first organism to produce mutated descendants do so, and over how many generations?

365. How many of those mutated descendants, if any, died without reproducing?

366. How many, if any, survived and reproduced?

367. What was the second organism, if any, to produce mutated descendants?

368. Was it the same species as the first organism to produce mutated descendants?

369. If not, what species was it?

370. How many times did the second organism to produce mutated descendants do so, and over how many generations?

371. How many of those mutated descendants, if any, died out without reproducing?

372. How many, if any, survived and reproduced?

373. If any of those mutated descendants survived and reproduced, what were they?

374.   Were they single-celled, multicellular, something less than a cell or some combination thereof?

375.   Which and how many, if any, were slight variations of their parent(s)?

376.   Which and how many, if any, were entirely new species?

377.   **What was the first evolved species to come into existence (i.e., the first descendant that was not the same species as its ancestors)?**

378.   Was it single-celled, multicellular or something less than a cell?

379.   When did the first evolved species come into existence?

380.   Where did the first evolved species come into existence?

381.   How did the first evolved species come into existence?

382.   How many unsuccessful and successful mutations occurred prior to the formation of the first evolved species?

383.   Of those mutations, which ones were essential to the formation of the first evolved species?

384.   Did only one mutation produce the first evolved species?

385.   Or were multiple mutations involved?

386. If multiple mutations were involved, did they occur one at a time, in series from one generation to the next?

387. Or did they occur simultaneously in a single offspring?

388. Or did they occur both in series and simultaneously?

389. Did the first evolved species successfully thrive and reproduce?

390. Or did it die out before reproducing?

391. How many unsuccessful evolved species came into existence, if any, before the first successful evolved species came into existence?

392. How many unsuccessful species, in total, have come into existence and died out?

393. What was the first *successful* evolved species to come into existence?

394. Was it a single-celled organism, a multicellular organism or something less than a cell?

395. When did the first successful evolved species come into existence?

396. Where did the first successful evolved species come into existence?

397. How did the first successful evolved species come into existence?

398. How much time passed between the formation of the first living organism and the formation of the first successful evolved species?

399.   If the first successful evolved species was not a descendant of the very first living organism, then how much time passed between (1) the formation of the first organism that would eventually have a descendant of a different species, and (2) the formation of that new species?

400.   In general, how many mutations does it take for one species to evolve from another?

401.   Does it vary from one case to another?

402.   If so, what is the minimum number of mutations necessary for one species to evolve from another?

403.   Also, is there a maximum number of mutations that can occur before a new species evolves from another?

404.   In general, how much time does it take for one species to evolve from another?

405.   Does it vary from one case to another?

406.   If so, what is the minimum amount of time necessary for one species to evolve from another?

407.   Can a single species exist without ever producing a new species, either because it never mutates or because it is somehow able to mutate indefinitely without producing a new species?

408.   If not, what is the maximum amount of time that a species can exist before mutations necessarily cause a new species to evolve from it?

409.   Which species, if any, have ever existed without ever mutating (regardless of whether they are descended

from an original, independent organism, as mentioned above)?

410.   How many such species, if any, have ever existed?

411.   Do any such species exist today?

412.   If so, which ones?

413.   Which species, if any, have ever existed that can mutate indefinitely without ever producing a new species?

414.   How many such species, if any, have ever existed?

415.   Do any such species exist today?

416.   If so, which ones?

417.   How many different species, of all kinds, exist today?

418.   How many different species have ever existed, including those that exist today and those that are now extinct?

419.   In general, if the existence and variety of life on Earth today are the result of a few billion years of random chance physical mutations, and chance events or conditions in nature (so-called "natural selection"), then how many random chance physical mutations have occurred, in total, since the first living organism came into existence?

420.   How many of those mutations, or characteristics arising from mutations, were successfully passed on to future generations?

421.   How many were not?

422.   More broadly, if Evolutionists don't know how long it took for the first DNA and the first amino acids to

form, and Evolutionists don't know how long it took for amino acids to form the first proteins, and Evolutionists don't know how long it took for DNA, proteins and other substances to form the first cell or the first living organism, and Evolutionists don't know how long it took for the first successful mutation to occur, and Evolutionists don't know how long it took or how many mutations it took for the first evolved species to come into existence, and Evolutionists don't know how many mutations have occurred in total up to now or how much time was required, then Evolutionists don't know, therefore, how much time the "descent of man" theory actually requires, so how can Evolutionists, or anyone else, know with scientific certainty if the Earth is old enough for the "descent of man" theory to be valid?

INSTINCT

423. **What is instinct?**

424. Is instinct, as one dictionary defines it, "the innate aspect of behavior that is unlearned, complex and normally adaptive"?

425. Does innate behavior require innate *knowledge* (i.e., knowledge possessed at birth)?

426. What are the differences between instinct, innate knowledge and involuntary bodily functions (i.e., bodily functions that don't require conscious thought to be carried out)?

427. Are there different types of instinct, some of which require innate knowledge and some of which do not?

428. How does innate knowledge come into existence?

429. In general, can random chance physical mutations generate the kinds of innate knowledge needed to produce the distinct behavioral characteristics found in various species?

430. If so, how?

431.   Can random chance physical mutations generate a way for such innate knowledge to be passed on from one generation to the next?

432.   If so, how?

433.   **For example, how does a spider know how to spin a web, given that spider offspring are not taught by their parents how to do so?**

434.   When did the first spider come into existence?

435.   Where did the first spider come into existence?

436.   How did the first spider come into existence?

437.   Did the first spider spin any webs?

438.   If so, how did the spider know how to do it?

439.   If not, what was the first spider to spin a web?

440.   When did it spin the first spider web?

441.   Where did it spin the first spider web?

442.   How did it know how to spin the web?

443.   How did it know it was supposed to spin a web?

444.   That is, why was the first spider to spin a web inclined to spin a web in the first place?

445.   Is it true that web-spinning spiders can secrete both "dry" and "sticky" filament?

446.   How do spiders know when to secrete dry filament and when to secrete sticky filament?

447.   How many random chance physical mutations were necessary to give the first spider to spin a web the necessary physical attributes to do so?

448.   What was the predecessor of the first spider to spin a web?

449.   Was it a spider that did not spin webs?

450.   If so, why didn't it spin webs?

451.   Did it lack a spinneret and the other physical attributes necessary to do so?

452.   Or did it have the physical capability to spin webs but not the knowledge or inclination to do so?

453.   Even if a series of random chance physical mutations over millions of years produced the physical spinnerets and bodily functions necessary for spiders to produce and secrete both dry and sticky filament, could random chance physical mutations have given the first spider to spin a web not only the knowledge to spin webs but the inclination to do so?

454.   If so, how?

455.   How are the knowledge and inclination to spin webs passed on from one generation of spiders to the next?

456.   Do spiders ever spin themselves into cocoons, either deliberately or accidentally?

457.   Why does the idea of an inept spider accidentally spinning itself into a cocoon seem so absurd?

458.   Does instinct necessarily preclude ineptitude?

459.   If so, why?

460.   **Isn't it equally absurd that some organisms deliberately spin themselves into cocoons?**

461.   Does a caterpillar do so, for example, because it has foreknowledge that it's going to undergo metamorphosis?

462.   If so, how could it possibly have such foreknowledge?

463.   If not, then why would it stop eating and spin itself into a cocoon?

464.   How does a caterpillar know *when* to stop eating and cocoon itself?

465.   Why don't caterpillars spin webs like spiders?

466.   Also, which came first, the butterfly or the butterfly caterpillar?

467.   From what previous species did butterflies and butterfly caterpillars evolve?

468.   Why do some caterpillars, like bagworms (a type of moth larvae), spin partial cocoons (or "bags") *before* they're ready to pupate?

469. And why, unlike other caterpillars, do bagworms put twigs and leaves in their cocoons?

470. Are the twigs and leaves used as camouflage?

471. Or are they added to turn the cocoon into a type of nest?

472. Or are they added for some other reason, or for a combination of reasons?

473. Regardless of why the twigs and leaves are added, when did the first bagworm to add twigs and leaves to its cocoon come into existence?

474. Where did the first bagworm to add twigs and leaves to its cocoon come into existence?

475. How did the first bagworm to add twigs and leaves to its cocoon come into existence?

476. What caused that particular bagworm to add the twigs and leaves to its cocoon?

477. How were the knowledge and inclination to add twigs and leaves to the cocoon passed on from one generation to the next?

478. More generally, what was the very first organism to spin itself into a cocoon and undergo a radical metamorphosis?

479. When did the first organism to spin itself into a cocoon and undergo metamorphosis do so?

480. Where did the first organism to spin itself into a cocoon and undergo metamorphosis do so?

481.   Where did the innate knowledge or inclination to cocoon itself come from, given that its predecessor did not cocoon itself?

482.   Even if a series of random chance physical mutations over millions of years produced the physical organs and bodily functions necessary for caterpillars to produce and secrete silky filament, could random chance physical mutations have given the first caterpillar to spin itself into a cocoon not only the knowledge to properly spin a cocoon but the inclination to do so at just the right stage in its development?

483.   If so, how?

484.   How are the knowledge and inclination to spin themselves into cocoons, at just the right stage of

development in their lives, passed on from one generation of caterpillars to the next?

485.   How could random chance physical mutations produce an organism so complex that it not only undergoes the radical physical transformation of metamorphosis, but does so on a delayed time schedule?

486.   **Speaking of delays, what was the first type of cicada to come into existence?**

487.   When did the first cicada come into existence?

488.   Where did the first cicada come into existence?

489.   How did the first cicada come into existence?

490.   Did the first cicada's offspring burrow into the ground after hatching?

491.   If so, how long did they remain in the ground before emerging to molt and mate?

492.   What eventually caused different types of cicadas to remain in the ground for different periods of time?

493.   For example, how do 17-year cicada "nymphs" know to remain in the ground for 17 years, while other types of cicadas remain in the ground for shorter periods?

494.   Why don't nymphs just stay underground and continue feeding?

495.   Do they somehow know in advance that they're going to molt and develop wings?

496.   How do they keep track of the passage of time?

497.   How do they know when to emerge?

498.   Do nymphs ever emerge too soon or too late?

499.   How are the knowledge and inclination to remain in the ground for 17 years, but then emerge to molt and mate, passed on from one generation to the next?

500.   Why, reportedly, do adult female cicadas insert their eggs into small tree branches, using a saw-like appendage to cut into the branches?

501.   Why don't female cicadas instead lay their eggs directly in the ground, so that when the young nymphs hatch they don't have to fall to the ground and burrow into it?

502.   Did the very first female cicada insert her eggs into small tree branches?

503.   If not, how did the first female cicada lay her eggs?

504.   When did the first female cicada to cut open a small tree branch and deposit her eggs inside do so?

505.   Where did the first female cicada to cut open a small tree branch and deposit her eggs inside do so?

506.   Why did the first female cicada to cut open a small tree branch and deposit her eggs inside do so, if her predecessor did not?

507.   From what predecessor organism did this female evolve and how did the predecessor lay its eggs?

508.   Did the predecessor also have a saw-like appendage, but without the knowledge or inclination to use it?

509.   If the predecessor organism didn't have a saw-like appendage, did the random chance physical

mutation(s) that produced the saw-like appendage in the descendant come with some sort of instruction manual?

510. That is, how did the first female cicada with a saw-like appendage know what to do with it?

511. Furthermore, even if a series of random chance physical mutations over millions of years produced the saw-like appendage on female cicadas that allows the females to insert their eggs into small tree branches, how do the females of each new generation acquire the knowledge and inclination to do so?

512. **What, for another example, was the first luminescent beetle?**

513. When did it come into existence?

514. Where did it come into existence?

515. How did it come into existence?

516. From what predecessor species did the first luminescent beetle evolve?

517. How many random chance mutations were needed to produce the first luminescent beetle's bioluminescent organ?

518. How long did those mutations take to produce that organ?

519. How did random chance physical mutations of the nonluminescent ancestor generate a beetle capable of producing the special chemicals needed for bioluminescence?

520.   Even if random chance physical mutations were able to produce a beetle with a bioluminescent organ, how do such beetles know to use such organs?

521.   For example, how do female fireflies (or lightning bugs) know to use their bioluminescence to attract males for mating?

522.   How do the males know to respond?

523.   In other words, why don't their bioluminescent organs just flash randomly and without any meaning?

524.   Did the very first fireflies communicate with their bioluminescent organs?

525.   If not, how many generations passed before fireflies began communicating with their bioluminescent organs?

526.   What triggered the change?

527.   How are the firefly's ability and inclination to use its bioluminescent organ to communicate passed on from one generation to the next?

528.   Besides bioluminescent beetles, what other organisms have bioluminescent capabilities?

529.   What was the very first organism to have a bioluminescent capability?

530.   When did it come into existence?

531.   Where did it come into existence?

532.   How did it come into existence?

533.   What was the second organism to have a bioluminescent capability?

534.   Was it a mutated descendant of the first bioluminescent organism or did its bioluminescent capability develop independently?

535.   How many times have random chance physical mutations independently produced bioluminescent organisms?

536.   What are the odds that random chance physical mutations would produce a single bioluminescent organism?

537.   What are the odds that more than one species of bioluminescent organisms (from bacteria to fireflies to sea creatures) would come into existence independently of each other, at different times and in different locations, with different methods for producing bioluminescence, each via a separate series of random chance physical mutations?

538.   **What was the first fish capable of emitting electrical discharges?**

539.   Was it an electric "eel" (so-called)?

540.   Was it an electric catfish?

541.   Was it an electric ray?

542.   Or was it a predecessor of one or more of these electric fish?

543.   When did the first electric fish come into existence?

544.   Where did the first electric fish come into existence?

545.   From what predecessor species did the first electric fish evolve?

546.   How many random chance mutations were needed to produce the first electric fish's ability to emit electrical discharges?

547.   How long did those mutations take to produce that capability?

548.   How, specifically, did random chance physical mutations of the nonelectrical predecessor species generate a fish capable of emitting electrical discharges?

549.   Even if random chance physical mutations were able to produce the physical attributes necessary for electrical discharges, how do electric fish know how to use such a capability—whether they use it to stun prey, to defend against predators or for other purposes?

550.   How are an electric fish's knowledge and inclination to use its electrical discharge capability passed on from one generation to the next?

551.   What are the odds that random chance physical mutations would produce a single electric fish?

552.   If electric eels, electric catfish and electric rays did not all evolve from the same electric predecessor fish, then how many times have random chance physical mutations independently produced electric fish?

553.   What are the odds that more than one species of electric fish would come into existence independently of each other, at different times and in different locations, with different methods for

producing electrical discharges, each via a separate series of random chance physical mutations?

554.   **What were the first fish to return to their river of origin to spawn?**

555.   When did those fish come into existence?

556.   Where did those fish come into existence?

557.   Why did they return to their river of origin to spawn?

558.   What was the predecessor species of fish?

559.   Why didn't the fish of the predecessor species return to their river of origin to spawn?

560.   Did only one fish, initially, return to its river of origin to spawn, only to discover that it takes more than one to tango?

561.   If so, did that fish fail to reproduce and die out?

562.   Or did that fish go back out to sea and return with other fish of the same species the following spawning season?

563.   If so, what caused other fish of that species to eventually return to the same river to spawn?

564.   Did the initial fish to do so communicate with other fish and convince them to join in?

565.   Indeed, what caused the initial fish to initially return?

566.   Or was the very first fish spawning migration a collective effort from the start?

567.   If so, what caused multiple fish of the same species and from the same river of origin to return to that river collectively, at roughly the same time, to spawn?

568.   How did they know when to return to spawn?

569.   How did they know how to get back to their river of origin to spawn?

570.   Why do some fish, like the Pacific salmon, reportedly return only once to spawn and then die?

571.   Why do other fish, like the Atlantic salmon, reportedly return to the same river repeatedly to spawn again and again?

572.   What caused this difference in spawning between the Pacific and Atlantic salmon?

573.   When did the difference in spawning between the Pacific and Atlantic salmon originate?

574. More generally, could random chance physical mutations have given various species of fish the knowledge and inclination to collectively return to their respective rivers of origin to spawn?

575. If so, how?

576. How are the knowledge and inclination to return to their river of origin to spawn passed on from one generation of fish to the next?

577. **What was the first beaver to build a dam?**

578. When did that beaver come into existence?

579. Where did that beaver come into existence?

580. How did that beaver come into existence?

581. When did it build its first dam?

582. Where did it build its first dam?

583. How did it know it should build a dam in the first place?

584. How did it know how to build a dam?

585. Did it learn through trial and error?

586. Or were its ability and inclination for dam building innate?

587. If innate, how did that innate knowledge come into existence?

588. Could random chance physical mutations generate such innate knowledge?

589. How was such knowledge passed on to future generations?

590. Why didn't the first dam-building beaver's predecessor organism build dams?

591. Which came into existence first: the beaver's inclination to fell trees and build dams, or its physical attributes that make it possible for the beaver to fell trees and build dams (e.g., its teeth and flat tail)?

592. Do individual beavers not only live together in beaver colonies but also work together, collectively, to build dams and lodges?

593. What were the very first beavers to work together in a colony?

594. When did the first colony of beavers to work together come into existence?

595. Where did the first colony of beavers to work together come into existence?

596. Were there predecessor beavers of the same or different species that did not work together in colonies?

597. If so, why didn't the predecessor beavers work together?

598. What caused the first colony of beavers to work together to do so, given that their predecessors did not work together?

599. That is, how did the first colony of beavers to work together *know* that they should work together?

600. Did they consciously decide to work together?

601.   How did the individual members of that first colony of beavers know *how* to work together, without duplicating each other's efforts?

602.   Did they have to learn by trial and error?

603.   In general, are the inclination and ability of beavers to work together, collectively, innate?

604.   How do the inclination and ability of beavers to work together get passed on to future generations?

605.   **What was the first bird to build a nest?**

606.   When did it build its first nest?

607.   Where did it build its first nest?

608.   How did it know how to build a nest?

609.   How did it know it should build a nest in the first place?

610.   Did it learn through trial and error?

611.   Or were its ability and inclination for nesting innate?

612.   If innate, how did that innate knowledge come into existence?

613.   Could random chance physical mutations generate such innate knowledge?

614.   How was such knowledge passed on to future generations?

615.   Did birds evolve from dinosaurs and, therefore, acquire their nesting skills and inclinations from their reptilian predecessors?

616.   If so, what was the first dinosaur or reptile to build a nest?

617.   When did it do so?

618.   Where did it do so?

619.   How did it know how to build a nest?

620.   How did it know it should build a nest in the first place?

621.   Did it learn through trial and error?

622.   Or were its ability and inclination for nesting innate?

623.   If innate, how did that innate knowledge come into existence?

624.   How was such knowledge passed on to future generations?

625.   Did all dinosaurs build nests?

626.   If not, why not?

627.   If there were nest-building dinosaurs, did any of them migrate?

628.   If so, did birds evolve only from nest-building, migrating dinosaurs?

629.   Or have some paleontologists gone too far and incorrectly assumed that dinosaurs had bird-like habits?

630.   **Speaking of migration, what was the first bird to make a seasonal migration?**

631.   How did it know when to migrate?

632.   How did it know where to go?

633.   How did it know it should migrate in the first place?

634.   How was the inclination to migrate passed on to future generations?

635.   When did the first bird migration occur?

636.   Where did the first bird migration occur?

637.   For example, when did birds in the northern hemisphere first fly south for the winter?

638.   Did they fly back the next season?

639.   If not, which birds were the first to fly south for the winter and then return the next season?

640.   Did random chance physical mutations give them genes for both wanderlust *and* homesickness?

641.   **What were the first birds to fly in an aerodynamic "V" formation?**

642.   When did they first fly in a V formation?

643.   Where did they first fly in a V formation?

644.   Why did they first fly in a V formation?

645.   Did they know that flying in a V formation was more efficient during long flights than flying in a haphazard grouping?

646.   If so, how did they know?

647.   Did they fly along the same route repeatedly, each time in a different formation, and keep track of the time it took for each trip, accounting for other variables, such as wind speed and wind direction?

648.   Or is that a ridiculous question?

649.   If the birds didn't know that flying in a V formation was more efficient during long flights, then why did they not only begin flying in a V formation but continue to do so?

650.   How does the inclination to fly in a V formation get passed on from one generation to the next?

651.   How could random chance physical mutations or so-called "natural selection" possibly give some types of birds, but not all, the inclination to fly in a V formation—not only on one occasion or even on random occasions, but routinely from one generation to the next—especially when the birds presumably don't have the necessary intelligence to know that flying in a V formation is sometimes, but not always, advantageous?

652.   In addition, how are some large flocks of birds able, as a group, to suddenly fly left, right, up and down, in a series of stunning synchronized movements, all with few or no collisions despite the individual birds being only inches apart from each other?

653.   What was the first flock to fly in such synchronized movements?

654.   When did that flock first fly in synchronized movements?

655.   Where did that flock first fly in synchronized movements?

656.   How did the ability for such synchronized flying first come into existence?

657.   How is that ability passed on from one generation to the next?

658.   How could random chance physical mutations or so-called "natural selection" possibly give some types of birds, but not all, the ability to fly in synchronization—not only on one occasion or even on random

occasions, but routinely from one generation to the next?

659.   Why do birds of a feather flock together in the first place?

660.   Did some random chance physical mutation create in them a gene for avian bigotry?

661.   **What was the first organism to fly?**

662.   Was it an insect, a reptile, a bird, a bat or something else?

663.   When did the first organism to fly come into existence?

664.   Where did the first organism to fly come into existence?

665.   How did the first organism to fly come into existence?

666.   Did the first organism to fly have wings?

667.   Was the first organism to have wings also the first organism to fly?

668.   What was the first organism to have wings?

669.   Was it an insect, a reptile, a bird, a bat or something else?

670.   When did the first organism to have wings come into existence?

671.   Where did the first organism to have wings come into existence?

672.   How did the first organism to have wings come into existence?

673.   Did the first organism to fly know, before it actually flew, that it had the necessary physical attributes to fly?

674.   Did it initially know how to use those physical attributes to fly?

675.   Or did it "crash" a few times before getting the hang of it?

676.   What was the predecessor organism of the first organism to fly?

677.   Why couldn't the predecessor fly?

678.   Did the predecessor have the wings or other necessary physical attributes, but not the inclination to attempt to fly?

679.   Or did the predecessor not have the necessary physical attributes to fly?

680.   How many random chance physical mutations were necessary to give the first organism to fly the physical ability to do so?

681.   How many random chance physical mutations were necessary to give the very first flying insect the wings and other necessary physical attributes it needed to be able to fly?

682.   How many random chance physical mutations were necessary to give the very first flying reptile the wings, lightweight bones or other physical attributes it needed to be able to fly?

683.   How many random chance physical mutations were necessary to give the very first flying bird the wings,

feathers, lightweight bones or other physical attributes it needed to be able to fly?

684.   How many random chance physical mutations were necessary to give the very first flying mammal the necessary physical attributes it needed to be able to fly?

685.   Did all flying organisms evolve from the same line of predecessors?

686.   Or is it highly unlikely, for example, that flying insects evolved from the same line of predecessors as, say, birds?

687.   Did multiple species of winged flying organisms evolve independently?

688.   What are the odds that a series of random chance physical mutations would result in a flying organism even once, when no flying organism had previously existed?

689.   What are the odds that more than one series of random chance physical mutations would each result in a flying organism independently of each other?

690.   **What was the first type of bee to come into existence?**

691.   When did the first bee come into existence?

692.   Where did the first bee come into existence?

693.   How did the first bee come into existence?

694.   Was it a queen, a drone, a worker or something else?

695.   Which came first, the bee or the grub?

696.   When did the first queen bee come into existence?

697.   Where did the first queen bee come into existence?

698.   How did the first queen bee come into existence?

699.   When did the first drone bee come into existence?

700.   Where did the first drone bee come into existence?

701.   How did the first drone bee come into existence?

702.   When did the first worker bee come into existence?

703.   Where did the first worker bee come into existence?

704.   How did the first worker bee come into existence?

705.   Did queens, drones and workers all come into existence simultaneously as a result of three separate series of parallel random chance physical mutations, all culminating at or about the same time, in three separate ancestral lines of predecessor organisms?

706.   If so, what are the odds of that?

707.   If not, how did the earliest bees survive and reproduce if they were not divided into queens, drones and workers as they are today?

708.   When did bees first become divided into queens, drones and workers?

709.   Where did bees first become divided into queens, drones and workers?

710.   How did bees first become divided into queens, drones and workers?

711.   When did bees establish the first hive?

712.   Where did bees establish the first hive?

713.   How did bees establish the first hive?

714.   Why did bees establish the first hive?

715.   Even if a series of random chance physical mutations could produce the complex physical differences that exist among just one species of bee, how could mere physical mutations cause bees to work together, collectively, in the complex ways that they do?

716.   How, for example, could mere random chance physical mutations have given the first bees to build a hive both the innate knowledge to build it and the inclination to collectively do so?

717.   How do the different types of bees know how to perform their respective specialized functions?

718.   Indeed, how does each individual bee know that it has a specialized role in the first place?

719.   Why doesn't each bee just go off on its own and fend for itself?

720. How are the knowledge and inclination to perform specialized functions *as part of a collective effort* passed on from one generation of bees to the next?

721. **Similarly, what was the first type of ant to come into existence?**

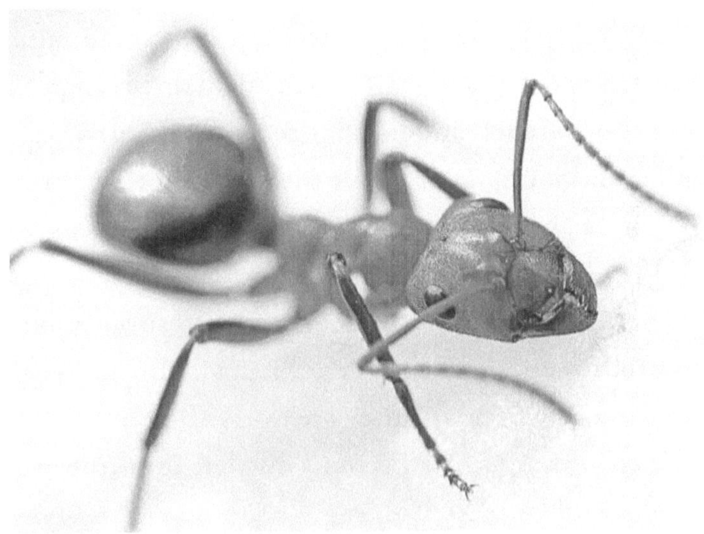

722. When did the first ant come into existence?

723. Where did the first ant come into existence?

724. How did the first ant come into existence?

725. Was the very first ant a queen, a male, a worker or something else?

726. Which came first, the ant or the ant larva?

727. When did the first queen ant come into existence?

728. Where did the first queen ant come into existence?

729. How did the first queen ant come into existence?

730.   When did the first male ant come into existence?

731.   Where did the first male ant come into existence?

732.   How did the first male ant come into existence?

733.   When did the first worker ant come into existence?

734.   Where did the first worker ant come into existence?

735.   How did the first worker ant come into existence?

736.   Did queens, males and workers all come into existence simultaneously as a result of three separate series of parallel random chance physical mutations, all culminating at or about the same time, in three separate ancestral lines of predecessor organisms?

737.   If so, what are the odds of that?

738.   If not, then how did the earliest ants survive and reproduce if they were not divided into queens, males and workers as they are today?

739.   When did ants first become divided into queens, males and workers?

740.   Where did ants first become divided into queens, males and workers?

741.   How did ants first become divided into queens, males and workers?

742.   When did ants establish the first ant nest?

743.   Where did ants establish the first ant nest?

744.   How did ants establish the first ant nest?

745.   Why did ants establish the first ant nest?

746.   Even if a series of random chance physical mutations over millions of years could produce the complex physical differences that exist among just one species

of ant, how could mere physical mutations cause ants to work together, collectively, in the complex ways that they do?

747. How, for example, could mere random chance physical mutations have given the first ants to build a nest both the innate knowledge to build it and the inclination to collectively do so?

748. How do the different types of ants know how to perform their respective specialized functions?

749. Indeed, how does each individual ant know that it has a specialized role in the first place?

750. Why doesn't each ant just go off on its own and fend for itself?

751. How are the knowledge and inclination to perform specialized functions *as part of a collective effort* passed on from one generation of ants to the next?

752. Did one series of random chance mutations produce bees with queens, drones (males) and workers, while a completely separate series of mutations produced ants with queens, males and workers?

753. If so, what are the odds that two such series of random chance mutations, each remarkable enough on its own to produce a tripartite (queen/male/ worker) insect species, would occur independently?

754. Which came into existence first, bees or ants?

755. If, however, bees and ants both evolved from the same tripartite predecessor species, then what was that predecessor species?

756.   When did it come into existence?

757.   Where did it come into existence?

758.   How did it come to be divided into queens, males and workers?

759.   Was it the very first tripartite insect species?

760.   If not, what was the very first tripartite insect species?

761.   When did it come into existence?

762.   Where did it come into existence?

763.   How did it come to be divided into queens, males and workers?

764.   How long did it take for the first tripartite insect species to evolve from its non-tripartite predecessor species?

765.   How many random chance mutations were necessary to produce the tripartite division?

766.   Did the queens, males and workers of the first tripartite insect species work together, collectively?

767.   If so, how did they acquire the knowledge and inclination to do so?

768.   If not, why not?

769.   If not, did the first tripartite insect species die out before reproducing because its members could not work together?

770.   If the first tripartite insect species died out before reproducing, then how many additional times did random chance mutations produce tripartite insect

species before one of those species survived and reproduced?

771.   How long did it take?

772.   If spiders are born with the innate knowledge to spin and use webs, bees are born with the innate knowledge to collectively build and use beehives, and ants are born with the innate knowledge to collectively build and use nests, then why aren't the children of mathematicians born with an innate knowledge of vector calculus and differential equations?

773.   **In general, does the concept of "instinct" actually *explain* the remarkable behaviors of insects and animals, including the examples given above, or is "instinct" just a clever ploy used by Evolutionists to *avoid* having to explain the origin and hereditary transmission of innate knowledge, inclination and collective behavior?**

774.   If the latter, would it be fair to say that the word "instinct" is one of the most successful propaganda terms ever invented?

775.   Is the vague concept of instinct any more rational, or any more plausible, than the idea that a monotheistic God has somehow preprogrammed his creatures with a variety of different behaviors and capabilities?

776.   Is the vague concept of instinct the only option available for people who deny the possibility of God?

SEX & STUFF

777. **What was the first species to reproduce sexually?**

778. How did *that* happen?

779. When did the first species to reproduce sexually come into existence?

780. Where did the first species to reproduce sexually come into existence?

781. When did the first organism to have male reproductive organs come into existence?

782. Where did the first organism to have male reproductive organs come into existence?

783. How did the first organism to have male reproductive organs come into existence?

784. When did the first organism to have female reproductive organs come into existence?

785. Where did the first organism to have female reproductive organs come into existence?

786. How did the first organism to have female reproductive organs come into existence?

787.   Were the first male and the first female of the same species?

788.   Or did the first male originate from one species and the first female originate from another species?

789.   Which came into existence first: an organism with male reproductive organs or an organism with female reproductive organs?

790.   Or did each come into existence simultaneously?

791.   If so, what are the odds of that?

792.   If not, then how did the first organism with male (or female) reproductive organs reproduce without a mate?

793.   Did the first organism with male (or female) reproductive organs also reproduce asexually?

794.   If so, how many asexually produced generations passed before a mate in the same species evolved with compatible reproductive organs of the opposite sex?

795.   That is, if a male evolved first, how long did it take for a compatible female to evolve, or vice versa?

796.   What are the odds that random chance physical mutations would produce an organism of one sex first and then, at some later point, produce an organism that's not only of the opposite sex but is of the same species and sexually compatible with the first organism?

797.   Did random chance physical mutations and so-called "natural selection," instead, produce a single,

hermaphroditic organism with both male and female reproductive organs before distinct male and female organisms had come into existence?

798. If so, what are the odds of that?

799. Did it reproduce successfully?

800. If so, how?

801. Did its descendants eventually evolve into separate male and female organisms?

802. If not, then how did separate male and female organisms originate?

803. If so, when did some descendants become only male?

804. Where did some descendants become only male?

805. How did some descendants become only male?

806. When did other descendants become only female?

807. Where did other descendants become only female?

808. How did other descendants become only female?

809. Did all of today's species that reproduce sexually evolve from the first species to reproduce sexually?

810. If not, how many times did sexual reproduction evolve independently?

811. For example, did sexual reproduction among plants originate independently of sexual reproduction among animals?

812. What are the odds that random chance physical mutations would independently produce two or more species that reproduce sexually?

813.   Also, did random chance physical mutations and so-called "natural selection" cause increasingly complex yet mutually compatible sex organs to develop simultaneously each and every time a species that reproduces sexually "evolved" from a predecessor species that reproduces sexually?

814.   In other words, if organisms that reproduce sexually have evolved over time from species that reproduce sexually with less complex sex organs into species that reproduce sexually with more complex sex organs, then what are the odds that random chance physical mutations would, for each such new species, *simultaneously* produce new male and new female sex organs that are compatible for reproduction, despite each having undergone such drastic mutation that they are sufficiently more complex to constitute an entirely new species—and therefore *not* capable of reproducing with the sex organs of respective predecessor species?

815.   And what are the odds that random chance physical mutations could cause this to happen not rarely, but over and over again for hundreds of millions of years, so that some primitive organism that reproduced sexually would eventually evolve into human beings with highly complex reproductive organs?

816.   Have mismatches ever occurred?

817.   That is, has any organism that reproduces sexually ever had a descendant that mutated in a way that

made its two sexes no longer compatible for reproduction?

818.   If so, did the mutated descendant become extinct because it couldn't reproduce?

819.   Or were the mutated male or female descendants able to reproduce with non-mutated descendants of the predecessor organism?

820.   Have any organisms that reproduce sexually ever had a descendant with one or more mutations that altered the descendant's male sex organ(s) without altering the descendant's female sex organ(s), or vice versa, *without* causing reproductive incompatibility?

821.   How many times, if ever, have such mismatches occurred?

822.   When did each occur?

823.   Where did each occur?

824.   In each case, what species was involved?

825.   By the way, is the natural male-to-female ratio for newborn humans almost one-to-one?

826.   If so, why?

827.   Wouldn't it be better for the survival of the species for the ratio to be one male for multiple females (say, one male for every three, five or ten females)—at least in the early years, before humans had overpopulated the planet?

828.   How could one or more random chance physical mutations have caused such a near-perfect one-to-one birth ratio?

829.   Does this one-to-one ratio at birth indicate that lifelong monogamy is a good idea for social stability?

830.   Have any species ever existed that had three or more sexes?

831.   If not, why not?

832.   That is, if asexual species evolved into sexual species with two sexes, and sexual species thrived because sexual reproduction provided greater genetic diversity or some other advantage to offspring, then why haven't species evolved with more than two sexes?

833.   **What was the first plant seed?**

834.   When did the first plant seed come into existence?

835.   Where did the first plant seed come into existence?

836.   How did the first plant seed come into existence?

837.   Which came first, a seed or a seed-bearing plant?

838.   That is, did some organism mutate directly into the first seed, or did a plant mutate and eventually produce the first seed?

839.   What was the immediate predecessor organism from which the first seed or the first plant to bear a seed evolved?

840.   How did that predecessor reproduce?

841.   Did the very first plant seed successfully germinate into a seed-producing plant?

842.   If not, why not?

843.   And what *was* the first seed to successfully germinate into a seed-producing plant?

844.   When did it come into existence?

845.   Where did it come into existence?

846.   How did it come into existence?

847.   What allowed it to succeed in germinating?

848.   Did all seed-producing plants evolve from the same original seed or from the same original seed-producing plant?

849.   Or have different seed-producing plants evolved independently of each other?

850.   If so, what are the odds of that?

851.   **What was the first organism to lay one or more eggs?**

852.   When did the first organism to lay eggs come into existence?

853.   Where did the first organism to lay eggs come into existence?

854.   How did the first organism to lay eggs come into existence?

855.   How did the predecessor species of the first species to lay eggs produce offspring?

856.   Did all egg-laying organisms evolve from the same original egg-laying organism?

857.   Or have different egg-laying organisms evolved independently of each other?

858.   If so, what are the odds of that?

859.   **What was the first organism to breathe air?**

860.   When did the first organism to breathe air come into existence?

861.   Where did the first organism to breathe air come into existence?

862.   How did the first organism to breathe air come into existence?

863.   Why didn't its predecessor species breathe air?

864.   Did all air-breathing organisms evolve from the same original air-breathing organism?

865.   If not, how many times have air-breathing organisms evolved independently?

866.   **Why do some organisms eat other organisms?**

867.   If some organisms, such as certain plants, can survive with only sunlight, air, water and nutrients from the ground—without eating any other organisms—then why would *any* organism eat another?

868.   What was the first organism to eat another organism?

869.   Was it single-celled, multicellular or something less than a cell?

870.   When did the first organism to eat another come into existence?

871.   Where did the first organism to eat another come into existence?

872.   How did the first organism to eat another come into existence?

873.   Did a random chance physical mutation, or a series of such mutations, result in an organism that eats other organisms?

874.   Did all of the aspects of eating—such as capturing, ingesting, digesting and excreting—develop simultaneously?

875.   If so, what are the odds of that?

876.   If not, then which aspects of eating developed first and which aspects of eating developed last?

877.   Over how many generations did these aspects develop?

878.   How, and at what point, did the first organism to eat another know it was supposed to eat other organisms, given that its predecessor did not eat other organisms?

879.   And what, exactly, did it eat?

880.   Did it eat only one kind of organism?

881.   If so, what kind did it eat?

882.   Or did it eat two or more different kinds of organisms?

883.   If so, which kinds?

884.   And how did it know what to eat?

885.   Did the first organism to eat another come into existence with a plentiful supply of food in its vicinity?

886.   If so, what are the odds of that?

887. If not, did the first organism to eat another become extinct because it died before it could reproduce?

888. Did organisms that eat others evolve more than once?

889. For example, did carnivorous plants evolve independently of carnivorous animals?

890. Did carnivorous insects, like spiders, evolve independently of carnivorous plants and other carnivorous animals?

891. If so, how many times did organisms that eat others evolve independently?

892. When did each such evolution occur?

893. Where did each such evolution occur?

894. How did each such evolution occur?

895. What are the odds that more than one independent series of random chance physical mutations would result in organisms that eat other organisms?

896. **What was the first organism to "mimic" another?**

897. When did the first organism to mimic another originate?

898. Where did the first organism to mimic another originate?

899. How did the first organism to mimic another originate?

900. Does mimicry really offer a survival advantage against predators?

901.   For example, are nonpoisonous butterflies with markings similar to those of poisonous butterflies really less likely to be eaten by birds?

902.   How would the birds know not to eat either type of butterfly?

903.   If a bird eats a poisonous butterfly and gets sick or dies, how do other birds come to realize that the problem was the butterfly?

904.   Does the sick or dying bird warn other birds?

905.   If so, how?

906.   Are birds, in general, able to communicate such specific information to other birds?

907.   Even if they are, how does the sick bird itself know that eating the poisonous butterfly made it sick?

908.   Do predators need to have some minimum level of intelligence for mimicry to be effective?

909.   Wouldn't mimicry also work in reverse, to the detriment of predators and butterflies alike?

910.   That is, wouldn't it be just as likely for birds to first encounter the nonpoisonous butterflies and safely eat them, and then later mistakenly eat the similarly marked poisonous ones?

911.   **How many organisms have symbiotic relationships with other organisms?**

912.   Is it plausible that the wide variety of symbiotic relationships among numerous organisms could

have developed through random chance physical mutations and so-called "natural selection"?

913.   When did the first symbiotic relationship come into existence?

914.   Where did the first symbiotic relationship come into existence?

915.   How did the first symbiotic relationship come into existence?

916.   Was the first symbiotic relationship *mutually beneficial* (like bees facilitating pollination when they get nectar from plants)?

917.   Or was it *parasitic* (in which one organism benefits at the expense of another)?

918.   What were the organisms involved in the first symbiotic relationship?

919.   Were they single-celled, multicellular, something less than a cell or some combination thereof?

920.   Did random chance physical mutations ever simultaneously generate symbiotic partner species?

921.   If so, what are the odds of that?

922.   For each case, if any, when, where and how have symbiotic partner species come into existence simultaneously?

923.   For each case, if any, which particular organisms were involved?

924.   Did symbiotic relationships ever develop over time (i.e., not simultaneously)?

925.   If so, what was the first symbiotic relationship to develop over time?

926.   Which organisms were involved?

927.   When did the symbiotic relationship begin?

928.   Where did the symbiotic relationship begin?

929.   How did the symbiotic relationship begin?

930.   In each case in which a symbiotic species depends on a symbiotic relationship for its very survival, how did the symbiotic species' nonsymbiotic predecessor species survive prior to the development of the symbiotic relationship?

931.   For example, were all parasites originally nonparasitic, but became parasitic through so-called "natural selection"?

932.   If so, then when, where and how did this occur for each parasite known to exist today?

933.   If not, did random chance physical mutations continuously give rise to various types of parasites over millions of years, including parasites that died immediately because they had no corresponding host organisms?

934.   If so, what are the odds that random chance physical mutations would generate such a wide variety of parasites that *did* have corresponding host organisms ready and waiting?

935.   **Also, what was the first virus?**

936.   When did the first virus come into existence?

937.   Where did the first virus come into existence?

938.   How did the first virus come into existence?

939.   Did the first virus "evolve," if that's possible, from some predecessor?

940.   If so, what was that predecessor?

941.   Was it a living organism that was more complex than a virus?

942.   Or was it nonliving organic matter that was less complex than a virus?

943.   Are viruses living organisms because they have genetic material, even if they can't replicate without a host?

944.   If not, does that mean that Evolution applies to nonliving things, given that viruses' genetic material is prone to mutation?

945.   Or does it mean that viruses can only mutate but never "evolve"?

946.   If viruses can't reproduce on their own, is it wrong or misleading to use the term "species" when classifying different types of viruses?

947.   Is it possible that a virus could mutate into a living organism capable of reproducing without a host organism?

948.   If so, has that ever happened?

949.   If so, when?

950.   If so, where?

951.   If so, how?

952.   **What is sleep?**

953.   Are there different kinds of sleep?

954.   What was the first organism to sleep?

955.   When did the first organism to sleep do so for the first time?

956.   Where did the first organism to sleep do so for the first time?

957.   Why didn't the predecessor organism sleep?

958.   How could random chance physical mutations produce an organism that has the need to sleep?

959.   How could a need for sleep be passed on from one generation to the next?

960.   How could random chance physical mutations have led to increasingly complex sleep functions, such as

those found in human beings and other mammals, as simpler species evolved into more complex ones?

961.   Which organisms do not sleep?

962.   Why don't they sleep?

963.   Are they all lower, invertebrate life forms?

964.   What is the most complex organism today that does not sleep?

965.   Have all organisms which do sleep evolved from the very first organism to sleep?

966.   If not, what are the odds that random chance physical mutations would independently produce more than one species that sleep?

967.   In general, why do any organisms sleep?

968.   For example, why sleep rather than just rest while fully conscious?

969.   What evolutionary advantage, if any, might sleeping convey to an organism?

970.   Would the disadvantages of sleeping, such as increased vulnerability to predators, offset any advantages?

971.   What is hibernation?

972.   Is hibernation a type of sleep?

973.   What was the first organism to hibernate?

974.   Where did the first organism to hibernate do so?

975.   When did the first organism to hibernate do so?

976.   Why didn't the predecessor organism hibernate?

977.   How could random chance physical mutations produce an organism that hibernates?

978. How could the need for, or habit of, hibernating be passed on from one generation to the next?

979. Have all organisms which hibernate evolved from the first organism to hibernate?

980. If not, what are the odds that random chance physical mutations would independently produce more than one species that hibernate in different ways, from insects to mammals?

981. **What is consciousness?**

982. What does it mean to be self-aware?

983. Are human beings the only living organisms that are self-aware?

984. If not, what other organisms today, if any, are self-aware?

985. What other organisms that are now extinct, if any, were self-aware?

986. What was the first living organism to have self-awareness?

987. When did the first organism to have self-awareness originate?

988. Where did the first organism to have self-awareness originate?

989. How did the first organism to have self-awareness originate?

990. From what predecessor did it evolve?

991. Why didn't the predecessor have self-awareness?

992.   How, specifically, could one or more random chance physical mutations give rise to consciousness or self-awareness?

993.   More generally, how could a universe originally devoid of consciousness produce organisms with consciousness?

994.   How are thoughts directed by consciousness?

995.   How is one thought distinguished from another?

996.   How and why do new ideas come into existence?

997.   How does consciousness distinguish new ideas from old ones?

998.   Are consciousness and self-awareness simply the result of large brains with sufficient neural activity?

999.   If so, does that mean inanimate computers will one day become self-aware if they are built with sufficient processing capacity?

1000.   Or is the idea of a conscious machine ludicrous?

1001.   Are consciousness and self-awareness, rather, indicators that a human being isn't simply a soulless organic machine, but a nexus of mental, physical and spiritual elements?

1002.   In other words, does consciousness or self-awareness require a spirit or soul of natural or supernatural origin?

1003.   Could Evolution, then, have a related ultimate purpose?

1004.   Is the ultimate purpose of Evolution to generate conscious, sentient beings, whose respective natural lives, in turn, prepare their spirits or souls for some "higher" existence after death?

1005. Or is it absurd to assert that billions of years of random chance physical mutations might have an ultimate purpose of any kind?

STACKED ASSUMPTIONS

1006. It's one thing to suggest that random chance physical mutations, followed by chance events or conditions in nature (so-called "natural selection"), might occasionally produce variations in the physical characteristics of organisms, but could *all* of the astounding complexity of life on Earth—more than a billion species, countless variations within each species, innate knowledge and inclination, collective cooperation, multifarious symbiotic relationships, and the existence of consciousness—be the result of nothing but random chance physical mutations, and chance events or conditions in nature, that began with a single, primitive organism?

1007. Why, for example, would random chance physical mutations lead to increasingly complex forms of life?

1008. Wouldn't truly random mutations be just as likely to produce less complex organisms as more complex organisms?

1009. Wouldn't truly random mutations also be just as likely to produce completely useless features that neither help nor harm an organism's chances for survival?

1010. If so, wouldn't such useless features be passed on from one generation to the next in organisms whose other features allow them to survive and reproduce?

1011. Shouldn't there now exist a wide variety of such organisms with a wide variety useless features?

1012. How many such organisms exist today?

1013. How many such useless features exist today?

1014. Furthermore, wouldn't truly random mutations be just as likely, or even more likely, to produce unsuccessful mutated organisms as successful ones—especially if success also depends on favorable chance events or chance conditions in nature?

1015. For example, are some or all of the millions of cases of cancer that occur each year caused by mutations?

1016. If so, wouldn't that mean that the vast majority of mutations are harmful, not helpful?

1017. Is it accurate, though, to assume that the existence of mutations is necessarily evidence of Evolution?

1018. Couldn't mutations exist in a world that was created—but later cursed with suffering and death—by a monotheistic God?

1019. Couldn't mutations exist in a world of immorality, suffering and death that has been inexplicably "manifested" by the cosmic consciousness of a pantheistic god-universe?

1020. Also, is it correct to assume that variations *within* each species are evidence of Evolution?

1021. Aren't variations within a species usually the result of flexibility within the complex realm of chromosomes, genes and DNA, rather than direct mutations?

1022. What originally caused the flexibility and epigenetic processes that allow chromosomes, genes and DNA to produce so many variations within so many species without direct mutations?

1023. Did chromosomes, genes and DNA come into existence via random chance combinations of organic molecules, or were they *and the information they contain* designed by God, or by highly intelligent space aliens, or by some other intelligent designer?

1024. In addition, how does an embryo grow from a single cell into a complex organism?

1025. That is, as the embryo's cells divide and multiply, how is it that some cells become one type of tissue or organ while other cells become another, such as skin cells, blood cells and liver cells?

1026. Do certain genes within a cell get activated or "turned on," while other genes are "turned off" in a process generally referred to as epigenetics?

1027. What was the very first organism with an epigenetic process?

1028.   When did it come into existence?

1029.   Where did it come into existence?

1030.   How did it come into existence?

1031.   Was the first organism with an epigenetic process single-celled?

1032.   Or are epigenetic processes only possible in multicellular organisms?

1033.   If so, what was the first multicellular organism?

1034.   When did it come into existence?

1035.   Where did it come into existence?

1036.   How many cells did it have?

1037.   Two, three or more than three?

1038.   What determined the number of cells?

1039.   Did the number of cells vary over the first multicellular organism's lifetime?

1040.   Why or why not?

1041.   More specifically, did the first multicellular organism begin life as a single cell and then split into two or more cells?

1042.   If not, what was the first multicellular organism to begin life as a single cell and then grow into two or more cells?

1043.   Was its predecessor a single-celled organism or a multicellular organism?

1044.   Did the first multicellular organism to begin life as a single cell and then split into two or more cells continue to split into more and more cells throughout its life until it died?

1045. Or did it eventually reach some maximum number of cells and stop growing?

1046. If so, what caused it to stop growing?

1047. And why did it stop at that particular point rather than some other point with a different maximum number of cells?

1048. If the first multicellular organism did not originate spontaneously out of nonliving matter, then what was its predecessor organism?

1049. Did the first multicellular organism have an epigenetic process?

1050. Or were all of its cells the same type throughout its life?

1051. What was the first multicellular organism to have different types of tissue?

1052. When did it come into existence?

1053. Where did it come into existence?

1054. How many different types of tissue did it have?

1055. Why didn't its predecessor have different types of tissue?

1056. Was the division of cells into different types of tissue caused by the same basic epigenetic processes that occur in many living organisms today?

1057. Or was such division caused by something else?

1058. If so, what?

1059. If the first multicellular organism with different types of tissue did not have an epigenetic process, then what was the first multicellular organism that did have an epigenetic process?

1060.   When did it come into existence?

1061.   Where did it come into existence?

1062.   Why didn't its predecessor have an epigenetic process?

1063.   How could one or more random chance physical mutations have caused the first epigenetic process to occur?

1064.   How could random chance physical mutations have caused increasingly complicated yet properly functioning epigenetic processes to occur in increasingly complicated organisms as species evolved?

1065.   How often, if ever, have epigenetic processes *not* functioned properly?

1066.   For example, have any organisms ever had skin cells grow inside of their livers, or liver cells grow inside of their brains, or toenails grow where eyelids should have been?

1067.   What are the odds that random chance physical mutations would eventually produce such a wide variety of species with highly complex epigentic processes without causing an innumerable number of horrifically deformed organisms?

1068.   How much time passed between the origin of the very first organism and the origin of the first multicellular organism with an epigenetic process?

1069.   During that time, how many mutations were required to produce the first multicellular organism with an epigenetic process?

1070.   Does Darwin's "descent of man" theory adequately account for the existence and incredible complexity of epigenetics?

1071.   That is, are random chance physical mutations really the best explanation for the existence not only of chromosomes, genes and DNA, but also of complex epigenetic processes that are necessary for complex organisms to exist?

1072.   If Darwin had known about DNA and the information it contains, and the reportedly incredible complexity of cells and epigenetic processes, would he have abandoned his theories about Evolution?

1073.   In general, how many random chance physical mutations, both successful and unsuccessful, have to have occurred, in total, to produce not only more than a billion different species, but all of the variations within each species, including all of the DNA, genes, chromosomes and other microscopic components of individual cells?

1074.   Trillions?

1075.   Quadrillions?

1076.   In addition, how many chance events and conditions in nature have to have occurred or existed *in conjunction* with the aforesaid random chance physical mutations for the "descent of man" theory of Evolution to be possible?

1077. Trillions?

1078. Quadrillions?

1079. **Can a theory really be scientific if it's based on an assumption of trillions or quadrillions of unobserved random chance physical mutations, and trillions or quadrillions of unobserved chance events and conditions?**

1080. Would it be more accurate to assert that each and every unobserved random chance physical mutation, and each and every unobserved chance event or condition in nature, is a mere assumption, not a scientific fact?

1081. If so, wouldn't that mean that Darwin's "descent of man" theory is not analogous to a stack of solid building blocks, where each block is an indisputable scientific fact, but is far more like a gigantic house of cards, in which many of the cards are wafer-thin assumptions about past events, stacked one on top of another, that can't be confirmed because no one can travel back in time to observe them?

1082. Has any new species ever been observed evolving from another, including each and every mutation involved?

1083. For example, given that fruit flies have very short life spans, allowing scientists to observe mutations in them over many generations, has anyone ever observed any fruit flies mutating into an entirely new and viable species?

1084.   If so, what was or were the new species?

1085.   If not, why not?

1086.   Even if the fossil record contains some samples that appear to support Evolution, does the fossil record clearly and indisputably reflect the gradual occurrence of trillions or quadrillions of random chance physical mutations over a few billion years?

1087.   Or does the fossil record contain only enough evidence to seemingly support the possibility that Evolution might sometimes occur, but too small a sampling of past organisms to confirm or refute Darwin's "descent of man" theory, given that only a relatively small percentage of past organisms were ever fossilized?

1088.   If the fossil record lacks sufficient evidence of transitional organisms, or so-called "intermediate forms," to fully support the "descent of man" theory, is it because the missing "intermediate forms" existed in the past but were never fossilized, or is it because the missing "intermediate forms" never existed in the first place?

1089.   Has the fossil record been distorted by one or more cataclysmic events in Earth's history?

1090.   For example, have there ever been global or near-global floods, such as the one mentioned in the Bible during Noah's day, that might have significantly altered sedimentary layers or other aspects of the fossil record?

1091. In any case, are Evolutionists' interpretations of the fossil record accurate?

1092. If the fossil record indicates, at some point, a relatively sudden and exponential increase in the variety and complexity of life on Earth, does that necessarily mean that DNA must be inherently prone to mutation, because mutations caused only by external triggers (such as radiation) would occur too infrequently to cause such an exponential increase?

1093. If most mutations *are* caused by inherently error-prone DNA, and if mutations really do increase the variety and complexity of living organisms over many generations, wouldn't that mean that: (A) in the early years of life on Earth, the type and number of mutations would be relatively low because the earliest organisms wouldn't be very complex and, hence, would have fewer potential opportunities and ways to mutate; but (B) as mutations themselves cause the variety and complexity of organisms on Earth to increase over time, there would be an exponential increase in both the number of mutations and the types of mutations, further increasing the variety and complexity of living organisms, because both the overall opportunity to mutate and the potential number of ways to mutate would be increasingly greater in a growing population of increasingly complex organisms?

1094. If so, that might explain how random chance physical mutations could lead to an overall increase in the complexity of organisms over time rather than a decrease, but why, then, haven't exponential increases in the number, variety and complexity of species continued to the present day?

1095. Do mutations *not* increase the variety and complexity of living organisms over time?

1096. Or is DNA *not* inherently prone to errors that cause mutations?

1097. If neither, then what caused the trillions or quadrillions of mutations that must have been necessary, according to the "descent of man" theory, to give us the variety and complexity of life on Earth today?

Alternatives

1098. **If the "descent of man" theory isn't true then, in the interest of fairness and open-mindedness, shouldn't alternative theories be mentioned, if only briefly, including those of a controversial or even dubious nature?**

1099. Is there, for example, some other explanation that better explains the variety and complexity of life on Earth, such as the theory of Intelligent Design?

1100. Is there any evidence that life on Earth might have been designed by highly intelligent extraterrestrials, or by one or more deities?

1101. Can design be inferred from the information in DNA, or from the reportedly remarkable processes that occur within a cell, or from some other aspect of living organisms?

1102. For example, is the theory of "irreducible complexity" valid?

1103. That is, do any organisms have physical mechanisms that are not merely complex, but are composed of components that would be useless on their own, individually, and thus would have to have come into existence simultaneously for the larger mechanism to work, *not* as the result of a gradual series of random chance physical mutations over time?

1104. If so, then how did those components come into existence?

1105. Is the theory of "irreducible complexity" invalidated, though, by the assumption that the features or functions of components of an organic mechanism don't change over time, from one generation to the next, as the host species mutates (prior to the formation of the organic mechanism in question)?

1106. Or is that merely a potential hazard that must be carefully avoided when searching for cases of irreducible complexity?

1107. Besides, such an assumption wouldn't apply to the very first cell and its components, would it?

1108. That is, wouldn't the very first cell in the very first living organism have to have been irreducibly complex because there were no previous generations in which mutations could have occurred and, therefore, the features and functions of the components of very first cell could not have evolved gradually?

1109. Wouldn't the very first DNA in that very first cell also have to have been irreducibly complex for the same reason?

1110. In other words, if:

(A) the very first living organism could not have been alive without at least one functioning cell;

(B) that very first cell, in turn, could not have been alive without functioning DNA and other components;

(C) that very first DNA, in turn, could not have functioned without multiple components of its own; and

(D) no previous living organisms existed from which the very first cell's components, including the very first DNA, could have evolved; then

(E) the components of the very first cell and the components of the very first DNA within that cell could not possibly have been the result of Evolution, right?

1111. If so, wouldn't that mean that the very first cell's components, including the very first DNA and its components, had to have been intelligently designed, given the improbability that various types of nonliving matter randomly combined multiple times in the same location not only to form the very first cell components and the very first DNA components, but in a way that resulted in all of those components successfully working together as a whole?

1112. Wouldn't that, in turn, necessarily and ironically mean that Darwinism itself requires one or more intelligent designers?

1113. Would any proven case of irreducible complexity be sufficient to refute the "descent of man" theory of Evolution?

1114. Or would it simply mean that there has been some other process at work in addition to Evolution?

1115. If so, then what other process?

1116. Also, is it important to recognize that the theory of Intelligent Design says nothing about the nature of the designer(s) and should therefore not be equated to Biblical Creationism because Intelligent Design could also be used to support non-Biblical theories of creationism?

1117. If so, would this mean that it's important to distinguish between (1) Biblical Creationism, which is based on the monotheistic God of the Bible and the book of Genesis, and is incompatible with the "descent of man" theory, and (2) non-Biblical theories of creationism, which could be based on intelligent extraterrestrials, polytheistic deities, a pantheistic deity or some other designer(s), any of which might or might not be compatible with the "descent of man" theory, depending on the specific details?

1118. **In addition to the possibility of Intelligent Design, are there other theories of Evolution that are *not* based on the notion that all life evolved from a single, original organism?**

1119. For example, is it possible that some species, but not all, are the result of occasional or sporadic Evolution?

1120. If so, then how would the non-evolved species have come into existence?

1121. Wouldn't any non-evolved organism have to have been intelligently designed?

1122. In other words, with or without sporadic Evolution, is some type of intelligent design the only other possible explanation for the existence and variety of life?

1123. Was life on Earth, for example, intelligently designed by highly advanced extraterrestrials?

1124. If so, then when, where and how did the extraterrestrials come into existence?

1125. Or has life on Earth been intelligently designed by polytheistic deities?

1126. If so, then when, where and how did the polytheistic deities come into existence?

1127. Could they all be eternal and, therefore, without origin?

1128. Or does plurality necessarily require a cause?

1129. Would extraterrestrials or polytheistic deities have limitations or shortcomings that would explain why they'd create flawed human beings and a world of immorality, suffering and death?

1130.   Or has life on Earth been intelligently designed and "manifested" by the cosmic consciousness of a pantheistic god-universe?

1131.   If so, then why are human nature and the world in general so inherently flawed?

1132.   Why wouldn't a pantheistic god-universe "manifest" people with better natures and a nicer world?

1133.   In other words, why would a pantheistic god-universe torment part of itself (i.e., humanity) with other parts of itself (e.g., natural disasters, diseases and dangerous animals)?

1134.   For pantheism to be true, wouldn't the cosmic consciousness of a pantheistic god-universe have to be a schizophrenic sadomasochist, given the widespread suffering in the world?

1135.   Would the existence of an eternal, monotheistic God make more sense than pantheism, polytheism or extraterrestrials?

1136.   After all, if the possibility of extraterrestrial life can't be ruled out because no one knows what might exist among the purportedly vast numbers of galaxies and assumed planets in the universe, then it would also be illogical, would it not, to rule out the possibility that God might exist somewhere within or outside of the universe?

1137.   Do Evolutionists assume that the very concept of an eternal God is inherently unscientific because they have an atheistic religious bias?

1138.   Wouldn't an eternal, monotheistic God provide a convenient, single solution to the various problems of ultimate origin, such as the origin of the universe and the origin of life?

1139.   Why, though, would God create sentient human beings?

1140.   Would God care about human beings, either collectively or individually?

1141.   Would God care about human beliefs or behavior?

1142.   In any case, why would a monotheistic God create a world of pain and suffering for his creatures, if not out of malice or insanity?

1143.   As the creator of gravity, electromagnetism and DNA, such a God couldn't be merely incompetent, right?

1144.   Is there any way to reconcile the possibility of a good, sane and competent God with such a bad world?

1145.   For example, could some version of Biblical Creationism and the Fall of Man be true, though perhaps more complicated than the sparsely detailed Biblical account reveals?

1146.   Is there any evidence that death might not have existed prior to the Fall of Man, at least for people and animals?

1147.   If so, is there any evidence that omnivores and carnivores did not exist prior to the Fall of Man?

1148. For any of the above to be true, wouldn't human and animal anatomy have to have been astoundingly different prior to the Fall?

1149. Have atheist Evolutionists—who seem to take too much pleasure when asking why men have nipples —simply failed to consider the possible effects of the Fall of Man on human and animal anatomy?

1150. Also, if people didn't have to toil prior to the Fall of Man, would that mean that entropy did not exist, or that the laws of physics were fundamentally different in some other way?

1151. More generally, if Biblical Creationism is true, shouldn't there be physical evidence to support it?

1152. Or would the effects of God's curse on anatomy, physical laws and other aspects of the natural world have eliminated such evidence, or somehow rendered it indiscernible to scientists today?

1153. Have Evolutionists ever seriously looked for evidence that doesn't fit their preconceived theories?

1154. Or do they merely assume that Evolution is true and only look for supporting evidence?

1155. If no one looks for contrary evidence, or if no one is allowed to look for it out of fear of being ridiculed, slandered, demonized, denied funding, blacklisted from publication, denied degrees or even entrance to schools, denied research or teaching positions, fired from their jobs or otherwise discriminated against, then Evolutionists can't claim to have withstood all challenges, can they?

1156. Ignoring, or even suppressing, opposing points of view isn't the same as refuting them, is it?

1157. Despite its usefulness for deniers of the supernatural, is Evolution compatible with the Bible and Christianity, as some have claimed?

1158. Or does one need to distinguish between Evolution in general and the "descent of man" theory in particular?

1159. That is, even if the God of the Bible might, theoretically, have created a world in which new species occasionally evolve from others, wouldn't the "descent of man" theory remain completely incompatible with the Biblical account of the Fall of Man?

1160. In other words, if the "descent of man" theory is true, then the Fall of Man never occurred, so wouldn't Christianity be pointless, because there would be no need for a Savior?

1161. What, after all, did Christ come to save people *from*, if not the negative consequences of the Fall of Man?

1162. Indeed, was Darwin's decision to title one of his books *The Descent of Man*[2] a deliberate play on words meant to highlight his religious opposition to the Biblical story of the *Fall* of Man?

---

[2]The full title is *The Descent of Man and Selection in Relation to Sex.*

ERROR AND DECEPTION

1163. **Are Evolutionists trustworthy?**

1164. How many times have Evolutionists made major mistakes?

1165. How many times have Evolutionists deliberately deceived people?

1166. Did the Brontosaurus, which I was taught about in school, ever really exist?

1167. Was the alleged error with the skull just a mistake or a lie?

1168. When a pig's tooth became the basis for Nebraska Man, was that a mistake or a lie?

1169. Was the alleged discovery of Piltdown Man a mistake or a lie?

1170. How long was the alleged Piltdown Man on display in a museum *after* it was known to be false?

1171. How many other Evolution-related museum displays have been faked?

Ernst Heinrich Haeckel (circa 1900)

1172. Was the "recapitulation" theory of Evolutionary
    embryology proposed by Ernst Heinrich Haeckel

(1834-1919), and depicted in his famous drawings of embryo development, a lie from the start?

1173.   Or was it just a theory that was later proven false?

1174.   Either way, why were copies of Haeckel's inaccurate drawings of embryo development still being displayed and taught in schools even when I was a child, decades after his theory was known to be false?

1175.   How often do Evolutionists falsify data to support their theories, as when Evolutionists used staged photos to support a theory that pollution had affected the evolution of peppered moths by darkening the bark of trees?

1176.   How many times have fake fossils been etched, carved, chiseled, sculpted or otherwise produced, such as that of the "Cardiff Giant," reportedly made in Cardiff, New York in the latter half of the 19th century?

1177.   Had the Cardiff Giant perpetrators been better sculptors, and had they chosen a scaled-up reptile skeleton for their subject rather than a humanoid figure, would they have pulled it off?

1178.   Have any fake fossils ever been produced in, say, China?

1179.   In general, how often have Evolutionists deliberately lied about their theories or alleged discoveries?

1180.   How often have Evolutionists lied to cover up their mistakes?

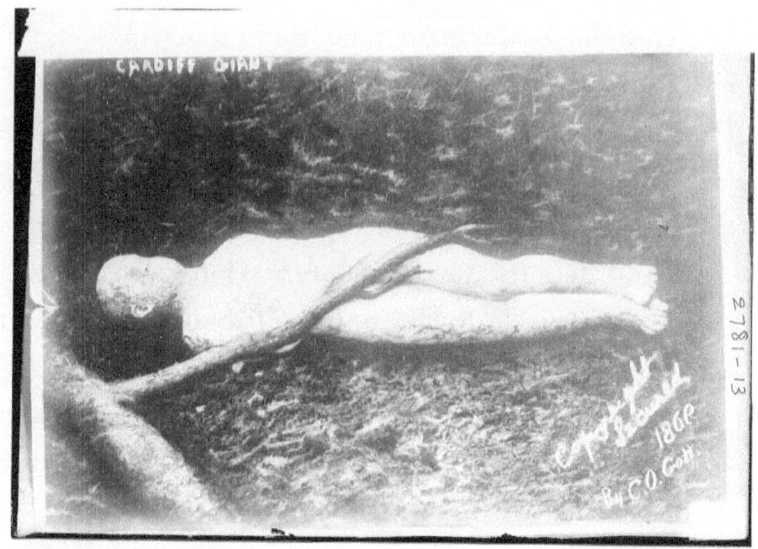

Library of Congress

1181. Given this history of repeated deceptions, in which I, personally, have been a victim, why should I believe anything that Evolutionists claim?

1182. When the theory of Evolution is taught to students, are any of these errors and lies mentioned?

1183. Are the assumptions and weaknesses of the theory taught as well as the strengths?

1184. How many of the questions I've listed here, if any, are ever presented to students?

1185. How many, if any, are ever answered?

1186. **If the majority of these questions can't be answered with demonstrable certainty, doesn't that mean that significant aspects of the "descent of man" theory of Evolution are founded not on scientific**

evidence—evidence based on experimentation and observation—but on sheer speculation and countless assumptions about an unobservable, untestable, unknowable past?

1187. Should any theory based on this amount of sheer speculation be considered scientific, much less true?

1188. Indeed, is Evolution just a theory, or set of theories, about biology?

1189. Or is Evolution also a fundamental religious tenet of atheist materialists and others who deny the supernatural, without which the deniers of the supernatural would be unable to explain the wide variety of life?

1190. Would it be wise to accept such a tenet without first questioning the assumptions upon which it's based?

1191. Has this religious aspect of Evolution led to unscientific bias among Evolutionists?

1192. Have Evolutionists ever misrepresented opposing points of view?

1193. For example, have Evolutionists ever misrepresented Biblical Creationism by ignoring the Fall of Man story and incorrectly assuming, instead, that a world created by a good and omnipotent God should be perfect, not "fallen" or cursed by God?

1194. That is, have Evolutionists ignored the possibility that mutations, deformities, diseases and death might, in whole or in part, be the result of God deliberately sabotaging certain aspects of his own handiwork as part of his curse on humanity and the world?

1195. Have Evolutionists ignored the possibility that similarities between different species, such as those between mammals, do not necessarily indicate Evolution from a common ancestor but might, instead, indicate that such species were designed by the same designer?

1196. Do Evolutionists use ridicule, slander and censorship to intimidate and silence people with opposing points of view, as I have heard?

1197. Does the use of such methods support or undermine the claims of atheist Evolutionists to have a monopoly on reason?

1198. Will I, a mere layman when it comes to biology, be ridiculed, slandered or censored by Evolutionist zealots for daring to raise questions that challenge one of their fundamental religious tenets?

PREJUDICIAL MAN

1199. **Was Darwin always honest, fair and unbiased?**

1200. For example, wouldn't it have been more honest, or at least more accurate, for Darwin to have titled his most famous book *The Origin of Species by Means of Countless Random Chance Mutations and Coincidental Environmental Conditions?*

1201. Or would a title like that have raised too many doubts about the theory?

1202. Is that why Evolutionists today tend to overemphasize so-called "natural selection" and downplay Darwinism's extraordinary reliance on random chance?

1203. Or did Darwin and his contemporaries simply not understand mutation and the critical role that it would have to play for his theory to be true?

1204. Also, did Darwin lie when he wrote, in Chapter XV of *The Origin of Species*, that "such cases as the presence of peculiar species of bats on oceanic islands and the absence of all other terrestrial mammals are facts utterly inexplicable on the theory of independent creation"?

1205. Utterly inexplicable?

1206. Why couldn't an omnipotent God create an island with bats on it but not other mammals?

1207. Or, if oceanic islands did not exist until after Noah's flood, why couldn't the animals on the island have gotten there in a way similar to Darwin's own explanation, though presumably over a shorter time period?

1208. In *The Origin of Species*, in Chapters XII and XIII, on the geographical distribution of species, why didn't Darwin at least mention Noah's flood?

1209. Was he afraid that his own theory regarding geographic distribution and oceanic islands might be used, at least in part, to support belief in the story of Noah's flood?

1210. Did Darwin lie when he wrote, in Chapter XV of *The Origin of Species*, "How inexplicable on the theory of creation is the occasional appearance of stripes on the shoulders and legs of the several species of horse genus and of their hybrids"?

1211. Why couldn't an omnipotent God have created horses or other species capable of variation?

1212. More generally, why couldn't variability within any particular species have been a part of God's original design?

1213. Or why couldn't variability within particular species be a result of God's curse on humanity after the Fall of Man?

1214. Why didn't Darwin acknowledge that theories of creation include the possibility that the similarities and differences between members of the same species are not evidence of Evolution from a more primitive common ancestor but, rather, are simply variations of an underlying design that was originally created by God to be variable or was later altered by God to be variable?

1215. Did Darwin redefine "creation" to mean something other than Biblical Creation and deliberately fail to tell his readers?

1216. Wouldn't that be just as deceptive as lying?

1217. Was Darwin's extraordinary verbosity deliberate obfuscation, intended to deter ordinary people from reading his books in their entirety or, at least, to help hide the weaknesses of his theory?

1218. Or was his prolix style simply an innocent result of what today might be called an obsessive-compulsive disorder?

1219. For example, even before writing *Origin of Species*, did Darwin spend eight years—*eight years!*—of his life writing a four-volume study of barnacles because he tended to obfuscate or because he was obsessive-compulsive?

1220. Or was it, perhaps, a little of both?

1221. Also, was Darwin sexist?

1222. Was Darwin correct when he wrote in *The Descent of Man*, Part III, Chapter XIX, that "man is more courageous, pugnacious and energetic than woman, and has a more inventive genius"?

1223. Even if Darwin was right about male pugnacity and inventiveness (ahem), do Evolutionists today believe that men are more courageous and energetic than women?

1224. In addition, was Darwin correct when he asserted, later in the same chapter, "thus, man has ultimately become superior to woman"?

1225. Should *that* be taught in schools?

1226. Do Evolutionists today believe that men are superior to women?

1227. Furthermore, was Darwin racist?

1228. For example, was Darwin correct when he wrote in *The Descent of Man*, Part I, Chapter VI, that "the break between man and his nearest allies will then be wider, for it will intervene between man in a more civilised state, as we may hope, even than the Caucasian, and some ape as low as a baboon, instead of as now between the negro (*sic*) or Australian and the gorilla"?

1229. Was Darwin asserting, in other words, that Caucasians are more evolved than blacks and indigenous Australians?

1230. Do Evolutionists today believe that whites are more evolved than blacks?

1231. Also, was it racist for Darwin to repeatedly use terms like "savages" and "barbarians"?

1232. Was Darwin correct when he wrote in *The Descent of Man*, Part I, Chapter VII, "the singular fact that the Europeans and Hindoos (*sic*), who belong to the same Aryan stock, and speak a language fundamentally the same, differ widely in appearance, whilst Europeans differ but little from Jews, who belong to the Semitic stock, and speak quite another language, has been accounted for by Broca, through certain Aryan branches having been largely crossed by indigenous tribes during their wide diffusion"?

1233. Were Darwin's assertions that (1) Europeans were Aryans but Jews were not; and (2) the Aryan race had become impure by breeding with other races; ever adopted by anyone else—say, in Germany?

1234. Is it unfair, though, to associate Darwin's "descent of man" theory with his personal beliefs and biases, especially if those biases merely reflected the prevailing views of the era in which he lived?

1235. Or is it important to consider the possibility that Darwin's assumptions and conclusions about Evolution might have been prejudiced, in whole or in part, by personal religious and possibly sexist and racist motivations?

Charles Darwin (1809-1882) (photo circa 1870)

1236. Was Darwin correct when he wrote in *The Descent of Man*, Part III, Chapter XX, "we have seen that idiots are often very hairy, and they are apt to revert in other characters to a lower animal type"?

1237. Did Darwin write that before or after growing his own long, bushy beard?

1238. Did he really believe—long before the advent of hippies—that some hairy people have "reverted" to a less evolved state?

1239. Do Evolutionists today believe there is a correlation between intellectual deficiency and physical hairiness?

1240. If not, then why are Evolutionists so willing to ignore such a bizarre and indefensible statement?

1241. If Christians or other proponents of intelligent design made such sexist, racist, bizarre and indefensible statements, would Evolutionists be equally willing to overlook them?

1242. In any case, hasn't Darwin's "descent of man" theory been fatally flawed from the very beginning because:

(1) Darwin, who had no knowledge of DNA or epigenetics, never fully explained the source of "variation," as he called it;

(2) Darwin never explained the origin of life, upon which his entire theory is based;

(3) Darwin never truly explained instinct or its origin; and

(4) Darwin never explained how much time, in total, Evolution would require or if the Earth was old enough to accommodate it?

1243. Didn't Darwin admit the first of these assertions when he wrote in *The Origin of Species*, Chapter V, "our ignorance of the laws of variation is profound"?

1244. Didn't Darwin admit the second when he wrote in Chapter XV of *The Origin of Species* that "…science as yet throws no light on the far higher problem of the essence or origin of life," and when he wrote in *The Descent of Man*, Part I, Chapter III, "in what manner the mental powers were first developed in the lowest organisms, is as hopeless an enquiry as how life itself first originated"?

1245. Didn't Darwin admit the third when he wrote in *The Origin of Species*, Chapter V, "I will not attempt any definition of instinct" and "many instincts are so wonderful that their development will probably appear to the reader a difficulty sufficient to overthrow my whole theory"?

1246. Didn't Darwin admit the fourth when he wrote in *The Origin of Species*, Chapter XI, "the foregoing objections hinge on the question whether we really know how old the world is, and at what period the various forms of life first appeared; and this may well be disputed"?

1247. **Can Darwinism be considered "settled science" when so many fundamental questions remain unanswered?**

1248. For example, do Evolutionists today know how life originated?

1249. Have they identified the very first organism to come into existence?

1250. Or the second, or the third, or the fourth?

1251. Do they know when, where and how the first organism originated?

1252. Or the second, or the third, or the fourth?

1253. Do they know when, where and how the first DNA originated?

1254. Or the second, or the third, or the fourth?

1255. Do they know when, where and how the first cell originated?

1256. Or the second, or the third, or the fourth?

1257. Do they know, with scientific certainty, that the very first cell and its components, including the very first

DNA and its components, were not intelligently designed?

1258. Do they know which organism was the first to successfully reproduce?

1259. Or the second, or the third, or the fourth?

1260. Do they know when, where and how those first reproductions occurred?

1261. Do they know which organism was the first to mutate?

1262. Do they know when, where and how the first mutation occurred?

1263. Or the second, or the third, or the fourth?

1264. Do they know what the results were of those early mutations?

1265. Have they identified the first species to evolve from another?

1266. Or the second, or the third, or the fourth?

1267. Do they know when, where and how that first evolved species came into existence?

1268. Or the second, or the third, or the fourth?

1269. Do they know how many mutations were required and how much time was required for the first evolved species to evolve?

1270. Or for the second, or the third or the fourth?

1271. Do they know if the first evolved species survived and successfully reproduced?

1272. Or the second, or the third, or the fourth?

1273. Have they identified the first multicellular organism?

1274. Or the second, or the third, or the fourth?

1275. Do they know when, where and how the first multicellular organism came into existence?

1276. Or the second, or the third, or the fourth?

1277. Do they know which multicellular organism was the first to have an epigenetic process that produced different types of cells (such as skin cells, blood cells and nerve cells)?

1278. Or the second, or the third, or the fourth?

1279. Do they know when, where, how and why the first epigenetic process occurred?

1280. Or the second, or the third, or the fourth?

1281. Have they identified the first species to reproduce sexually?

1282. Or the second, or the third, or the fourth?

1283. Do they know when, where and how the first sexual reproduction occurred?

1284. If Evolutionists today don't know the answers to these questions, isn't the "descent of man" theory just as fatally incomplete now as it was when Darwin first invented it?

1285. Once the questions in this list are taken into consideration, can anyone honestly claim that Evolution irrefutably explains, or even convincingly explains:

(1) the origin of life;

(2) the wide variety of life (including both the physical *and behavioral* characteristics of all living organisms); and

(3) the nature and existence of consciousness;

in a way that obviates the possibility of intelligent design?

———————————————

About the Author

Anyone who wants to know more about Mark Burton should read his book *The Most Important Question in the World* (Second Edition). It's a philosophical treatise about certainty, uncertainty and ultimate truth. It reveals him at his best, which is how he would prefer to be remembered.

INDEX